Rebecca Winkelmann

KLF2, a master regulator of peripheral B lymphocyte homeostasis

Rebecca Winkelmann

KLF2, a master regulator of peripheral B lymphocyte homeostasis

Investigations on the function of Krüppel-like factor 2 (KLF2) in B lymphocytes

Südwestdeutscher Verlag für Hochschulschriften

Imprint
Any brand names and product names mentioned in this book are subject to trademark, brand or patent protection and are trademarks or registered trademarks of their respective holders. The use of brand names, product names, common names, trade names, product descriptions etc. even without a particular marking in this work is in no way to be construed to mean that such names may be regarded as unrestricted in respect of trademark and brand protection legislation and could thus be used by anyone.

Publisher:
Südwestdeutscher Verlag für Hochschulschriften
is a trademark of
Dodo Books Indian Ocean Ltd., member of the OmniScriptum S.R.L Publishing group
str. A.Russo 15, of. 61, Chisinau-2068, Republic of Moldova Europe
Printed at: see last page
ISBN: 978-3-8381-2572-5

Zugl. / Approved by: Erlangen, FAU, Diss., 2010

Copyright © Rebecca Winkelmann
Copyright © 2011 Dodo Books Indian Ocean Ltd., member of the OmniScriptum S.R.L Publishing group

Table of contents

I.	Figures	4
II.	Tables	5
III.	Abbreviations	6

1 Introduction 11
   1.1 B cell development and homeostasis 11
      1.1.1 Central B cell development 11
      1.1.2 Peripheral B cell development 14
      1.1.3 Plasma cell generation and migration 17
   1.2 Krüppel-like factor 2 (KLF2) 19
      1.2.1 Krüppel-like transcription factors 19
      1.2.2 KLF2 in T lymphocytes 20
      1.2.3 Regulation of KLF2 21

2 Results 23
   2.1 B cell speficic KLF2 deletion *in vivo* 23
      2.1.1 KLF2 expression profile during B cell development 23
      2.1.2 Normal B cell development in the bone marrow of KLF2-deficient animals 25
      2.1.3 Enlarged spleen size and increased numbers of splenic B cell populations in KLF2-deficient animals 27
      2.1.4 Decreased frequencies of B cells in the peritoneal cavity, the blood and peyers patches of KLF2-deficient mice 31
      2.1.5 Reduced numbers of plasma cells in the bone marrow after boost immunization with TD antigen 35
      2.1.6 Distribution of peripheral B cell subsets is controlled by KLF2 regulating L-selectin and $\alpha_4\beta_7$ integrin but not S1P1 38
      2.1.7 KLF2 determines Fo B cell identity 42
   2.2 Biochemical analyses of KLF2 43
      2.2.1 KLF2 protein modifications in splenic B cells and non B cells 43
      2.2.2 *In silico* phosphorylation analysis of KLF2 amino acid sequence 45
      2.2.3 KLF2 is posttranslationally modified by phosphorylation 46

3	**Discussion**		**48**
3.1	KLF2-deficiency results in impaired B cells homeostasis and plasma cell homing		48
3.2	Biochemical analyses of KLF2		52
4	**Material and Methods**		**56**
4.1	Material		56
	4.1.1	Chemicals	56
	4.1.2	Antibodies	56
	4.1.3	PCR primer	59
	4.1.4	Plasmids	60
	4.1.5	Cell lines	61
	4.1.6	Mouse strains	61
	4.1.7	Bacteria	62
	4.1.8	Technical Equipment	62
	4.1.9	Plastic material	63
	4.1.10	Consumable material	64
4.2	Methods		65
	4.2.1	Cell culture	65
	4.2.1.1	Cultivation and harvest of vertebrate cell lines	65
	4.2.1.2	Cryoconservation	65
	4.2.1.3	Thawing of cryoconserved cells	65
	4.2.1.4	Cell counting	66
	4.2.2	Isolation of primary B and T lymphocytes from different organs	66
	4.2.2.1	Erythrocyte depletion	66
	4.2.2.2	MACS-sorting of CD43⁻ splenic and CD19⁺ bone marrow B cells	66
	4.2.3	Flow cytometry	67
	4.2.3.1	Analyses of surface expression and intracellular proteins	67
	4.2.3.2	MoFlo^{TM} cell sorting	67
	4.2.4	Histology	67
	4.2.5	Molecular biological methods	68
	4.2.5.1	Plasmid DNA and RNA preparation	68
	4.2.5.2	DNA restriction analyses	68
	4.2.5.3	DNA fragment agarose gel electrophoresis	68
	4.2.5.4	Isolation of genomic tail DNA	69
	4.2.5.5	DNA/RNA concentration	69

4.2.5.6	Polymerase chain reaction (PCR)	69
4.2.5.7	cDNA synthesis	70
4.2.5.8	Qualitative RT-PCR	70
4.2.5.9	Quantitative TaqMan®-RT-PCR	70
4.2.5.10	Affymetrix microarray analysis	71
4.2.6	Antibody isotype detection using ELISA	71
4.2.7	IgG specific Elispot	71
4.2.8	Proteinbiochemistry	72
4.2.8.1	Cell lysates	72
4.2.8.2	Nuclear extracts	72
4.2.8.3	BCA test	72
4.2.8.4	Phosphatase inhibitor and Ly294002 stimulation	72
4.2.8.5	Immunoprecipitation	73
4.2.8.6	SDS PAGE	73
4.2.8.7	Western blot analysis	74
4.2.9	Calcium measurements	74
4.2.10	Transfection and infection of vertebrate cells	75
4.2.10.1	Transient transfection of adherent cells using calcium phosphate	75
4.2.10.2	Infection of NIH3T3 cells	75
4.2.11	*In vivo* methods	75
4.2.11.1	Immunization	75
4.2.11.2	FTY720-treatment	76
4.2.11.3	BrdU-treatment	76
4.2.12	Statistics	76

5 Epilogue   77

6 References   78

Figures

Figure 1. Schematic view of B cell development. 13
Figure 2. Formation of plasma cells. 18
Figure 3. Role of KLF2 in programming T cell quiescence. 20
Figure 4. Analyses of KLF2 expression profile in different B cell lines. 23
Figure 5. Analyses of KLF2 expression profile in B cells. 24
Figure 6. Schematic view of KLF2 expression profile during B cell development. 25
Figure 7. KLF2 conditional knockout strategy. 26
Figure 8. Analysis of B cell precursors and recirculating B cells in the bone marrow from wildtype and KLF2-deficient mice. 27
Figure 9. Analysis of of spleen physiognomy from wildtype and KLF2-deficient mice. 28
Figure 10. Analysis of B cells in spleen from wildtype and KLF2-deficient mice. 29
Figure 11. KLF2 deficiency does not affect *in vitro* survival capacity and proliferation of Fo and MZ B cells. 30
Figure 12. Analysis of spleen architecture from wildtype and KLF2-deficient mice. 31
Figure 13. Analysis of B cells in lymph node, liver and blood from wildtype and KLF2-deficient mice. 32
Figure 14. Analysis of thymi from wildtype and KLF2-deficient mice. 33
Figure 15. Analysis of peyers patches from wildtype and KLF2-deficient mice. 33
Figure 16. Analysis of peritoneal B cell subsets from wildtype and KLF2-deficient mice. 34
Figure 17. Analysis of antibody titers in non-immunized wildtype and KLF2-deficient mice. 34
Figure 18. Analyses of Ig-titers after immunization of wildtype and KLF2-deficient mice with T-independent antigens. 35
Figure 19. Analyses of Ig-titers after immunization of wildtype and KLF2-deficient mice with the T-dependent antigen TNP-KLH. 36
Figure 20. Analyses of plasma cells after immunization of wildtype and KLF2-deficient mice with the T-dependent antigen TNP-KLH. 38
Figure 21. Identification of potential KLF2 target genes. 39
Figure 22. Verification of KLF2 target genes. 41
Figure 23. Analyses of complement receptor (CD21/35) expression on spleen and lymph node B cells from wildtype and KLF2-deficient mice. 43
Figure 24. KLF2 expression pattern after different sorting strategies and fractionizing into nuclear and cytosolic proteins. 44
Figure 25. *In silico* analysis of potential KLF2 phosphorylation sites. 45

Figure 26. KLF2 is posttranslationally modified by phosphorylation. 47
Figure 27. KLF2 phosphorylation via PI3K signalling leads to its degradation. 54

Tables

Table 1. Summary of transitional B cell surface marker expression   14
Table 2. Antibodies in alphabetical order   56
Table 3. Oligonucleotides   59
Table 4. Established plasmids   60
Table 5. Mouse strain overview   62
Table 6. Technical equipment in alphabetical order   62
Table 7. Plastic material in alphabetical order   63
Table 8. Consumable material in alphabetiacl order   64
Table 9. PCR component overwiew   69
Table 10. Overwiew of components for SDS PAGE   74

Abbreviations

A	adenine
α	alpha
AA	amino acid
AID	activation induced deaminase
APRIL	a proliferation-inducing ligand
ASC	antibody secreting cell
β	beta
Bach2	BTB and CNC homology 1
BAFF	B cell activating factor
BCAP	B-cell PI3K adaptor protein
BCL6	B-cell CLL/lymphoma 6
BCMA	B cell maturation antigen
BCR	B cell receptor
Blimp-1	B lymphocyte induced maturation protein
BrdU	5-bromo-2-deoxyuridine
BSA	bovine serum albumine
BTK	Bruton's tyrosine kinase
C	cytosine
°C	temperature in Grad Celsius
Ca^{2+}	calcium ions
CCR	C-C chemokine receptor type
CD	cluster of differentiation
cdk2	Cyclin-dependent kinase-1
CFA	complete Freud's adjuvants
Cl^{2-}	chloride ions
CLP	common lymphoid progenitor
CO_2	carbon dioxide
cre	cre recombinase
CSR	class switch recombination
CxCR	C-X-C chemokine receptor type
d	days
DL-1	Delta like-1
DMSO	dimethylsulfoxide
DNA	desoxyribonucleic acid

E2A	E2A immunoglobulin enhancer-binding factor E12/E47
EBF-1	early B-cell factor 1
EBV	Epstein-Barr-Virus
EDTA	ethylenediaminetetraacetic acid
ELISA	enzyme-linked immunosorbent assay
Erk1	mitogen-activated protein kinase 3
FACS	fluorescence activated cell sorting
FCS	fetal calf serum
Fo	follicular
Foxo	forkhead box O
γ	gamma
G	guanine
g	gram
Gadd45	Growth arrest and DNA damage- induced gene 45
GC	germinal center
h	hours
HBS	HEPES buffered saline
HEK	human embryonic kidney
HEVs	high endothelial venules
H_2O	water
HPRT	hypoxanthine-guanine phosphoribosyltransferase
HRP	horse-raddish-peroxidase
HSC	hematopoietic stem cell
iCFA	incomplete Freud's adjuvants
IF	immunofluorescence
Ig	immunoglobulin
IgH	immunoglobulin heavy chain
IgL	light chain
IL	interleukin
IP	immunoprecipitation
i. p.	intraperitoneally
IRF4	interferon regulatory factor 4
i. v.	intravenously
κ	kappa
K^+	potassium ions
KLF2/LKLF	Krüppel-like factor 2

KLH	keyhole limbet hemocyanin
KO	knockout
λ	lambda
L	ligand
l	liter
lck	lymphocyte-specific protein tyrosine kinase
LPS	lipopolysaccharide
m	meter
µ-	micro-
m-	milli-
M	molar
MACS	Magnetic Cell Separation
MAdCAM-1	Mucosal vascular addressin cell-adhesion molecule 1
Mg^{2+}	magnesium ions
µHC	µ heavy chain
min	minute
MINT	MSH-homeobox-homologue 2-interacting nuclear target
MOMA	sialic acid binding Ig-like lectin 1
MZ	marginal zone
n-	nano-
Na^+	sodium ions
NZB/W	New zeeland black/white
OD	optical density
OH	hydroxide
P	phospho-
Pax5	paired box protein 5
PBS	phosphate buffered saline
PCR	polymerase chain reaction
PCs	plasma cells
PI3K	phosphoinositide 3-kinase
PKB/Akt	Protein kinase B
PLCγ	Phospholipase Cγ
PP	peyers patches
pre-BCR	pre-B cell receptor
prepro	precursor progenitor
pro	progenitor

PtdInsP3	phosphatidylinositol-(3,4,5)-phosphate
PU.1	SFFV proviral integration 1
RAG	recombination activating gene
RNA	ribonucleic acid
rpm	rounds per minute
RT	room temperature
RT-PCR	reverse transcription polymerase chain reaction
SDS	sodium dodecyl sulfate
SEM	standard error
SLP-65	SH2-domain-containing leukocyte protein of 65 kDa
S1P	sphingosin 1-phosphate
S1P1	sphingosin 1-phosphate receptor 1
S1P3	sphingosin 1-phosphate receptor 3
Src	v-src sarcoma (Schmidt-Ruppin A-2) viral oncogene homolog
Syk	spleen tyrosine kinase
T	thymidine
T	transitional
TD	thymus-dependent
TI	thymus-independent
TI-1	thymus-independent antigens type 1
TI-2	thymus-independent antigens type 2
TNF	tumor necrosis factor
TNP	2,4,6-trinitrophenol
U	units
U	uracil
v	volume
vav	vav guanine nucleotide exchange factor
VCAM1	Vascular cell adhesion molecule 1
WB	western blot
WT	wildtype
WWP1	WW domain containing E3 ubiquitin protein ligase 1
XBP1	X box binding protein-1

1 Introduction

1.1 B cell development and homeostasis

B lymphocytes play a prominent role in the combat of infectious diseases. The effector B cells, called plasma cells (PCs) or antibody secreting cells (ASCs), secret antibodies which specifically bind, neutralize and eliminate pathogens or toxins by recruitment of phagocytotic cells.

Antibodies consist of two identical heavy and light immunoglobulin chains. Since antibodies produced by a single plasma cell can only detect one specific antigen, diversity is obtained during B cell development by gene segment (V, D, J-segment) rearrangements at the heavy and light chain locus and differentiation is controlled by several checkpoints.

1.1.1 Central B cell development

B cell development in vertebrates takes place in the bone marrow (Fig. 1). B lymphoid cells originate from a hematopoietic stem cell (HSC) along a highly ordered but flexible pathway. HSCs are pluripotent stem cells giving rise to all lymphoid (B cells, T cells, natural killer cells) and myeloid cells (erythrocytes, megacaryocytes). Development into B or T cells segregates after the common lymphoid progenitor (CLP) stage by upregulation of transcription factors like E2A, EBF-1, PU.1, Ikaros leading to the differentiation into precursor progenitor (prepro-) B cells (reviewed in Matthias and Rolink, 2005). In the next stage, the progenitor (pro-) B cell, D_H to J_H rearrangement is initiated at the immunoglobulin heavy chain (IgH) locus by upregulation of the recombination complex consisting mainly of Rag1 and Rag2 (recombination activating gene 1 and 2) followed by rearrangement of V_H to DJ_H in the late pro-B stage (Mombaerts et al., 1992; Shinkai et al., 1992; Tonegawa, 1983; Willerford et al., 1996). After productive rearrangement, two heavy chains (μHC) can assemble with the surrogate light chain components VpreB and $\lambda 5$ as well as the signalling molecules Igα (encoded by the mb1 gene) and Igβ, (encoded by the B29 gene) to form the pre-B cell receptor (Hombach et al., 1990; Karasuyama et al., 1990; Tsubata and Reth, 1990).

Expression of the pre-B cell receptor (pre-BCR) is part of a crucial checkpoint in early B cell development, where the newly generated μHC is tested for functionality. V(D)J recombination is often improper resulting in truncated, unpairing or non functional μHCs (reviewed in Tonegawa, 1983). Only pre-B cells expressing a functional μHC are allowed to differentiate further (Brouns et al., 1993; Hombach et al., 1990; Kline et al., 1998); others are eliminated by apoptosis (Fang et al., 1996; Kline et al., 1998; Lu and Osmond, 1997). Large pre-B cells start to proliferate and downregulate the surface marker c-kit and upregulate CD25 as well as CD19 (Faust et al., 1993; Rolink et al., 1994). Furthermore Rag1 and Rag2 are downregulated (Grawunder et al., 1995) to inhibit rearrangement at the second μHC locus. This process is called allelic exclusion and results in pre-B cells with only one rearranged μHC allele, expressing only one μHC which is presented on the surface (reviewed in Jung et al., 2006). The pre-BCR induces a burst of proliferative clonal expansion to compensate for the loss of non functional pre-B cells. Furthermore the clonal expansion elevates the repertoire because each μHC can pair with a different light chain in emerging daughter cells (Hess et al., 2001; Rolink et al., 2000). After 4-6 divisions pre-B cells differentiate further into small pre-B cells where rearrangement of the immunoglobulin light chain (IgL) locus is initiated. However, the molecular mechanisms underlying termination of pre-B cell proliferation are not fully understood. It was shown that transcription factors such as Ikaros and Aiolos can induce cell cycle arrest in pre-B cells (Ma et al., 2010; Ma et al., 2008; Thompson et al., 2007; Trageser et al., 2009). In addition, Foxo proteins can induce IgL gene recombination through activation of RAG and induction of a delay in the G1 phase of the cell cycle (Amin and Schlissel, 2008; Herzog et al., 2008). We could show that KLF2 is a late target gene of the pre-BCR (Schuh et al., 2008). KLF2 is a target gene of Foxo1 (Fabre et al., 2008) and can induce quiescence in T cells (Buckley et al., 2001; Haaland et al., 2005; Kuo et al., 1997; Wu and Lingrel, 2004).

Immunoglobulin light chains (IgLs) are encoded on two gene loci, the IgLκ and the IgLλ locus, but consist only of V and J segments. First rearrangement takes place at the IgLκ locus. Only if recombination was unproductive the segments at the IgLλ locus are rearranged (Geier and Schlissel, 2006; Langerak and van Dongen, 2006) predominantly resulting in B cells with a light chain of the κ-type with a typical ratio of 20:1 (McGuire and Vitetta, 1981).

After assembly of the B cell receptor (BCR) immature B cells pass two additional checkpoints where their BCR is tested for functionality and autoreactivity. Potentially autoreactive B cells or B cells expressing a non-functional μHC can be rescued by a mechanism called receptor editing (Nemazee and Weigert, 2000). This mechanism implies rearrangement of the second IgLκ or the IgLλ locus for production of a functional non autoreactive BCR (Edry and Melamed, 2004; Gay et al., 1993; Nemazee and Weigert, 2000; Radic et al., 1993; Tiegs et al., 1993; von Boehmer and Melchers, 2010) and is thought to be regulated by BCR signalling strength (reviewed in Tussiwand et al., 2009). Functionally, non autoreactive immature B cells are characterized by high expression of IgM and low expression of IgD (Loder et al., 1999) with a tonic BCR signal, switch of RAG expression and then leave the bone marrow by upregulation of sphingosin 1-phosphate receptor 1 (S1P1). Since concentration of the S1P1 ligand sphingosin 1-phosphate (S1P) is high in the blood, immature B cells are attracted to the blood and are destined for the spleen for further differentiation (Pereira et al., 2010).

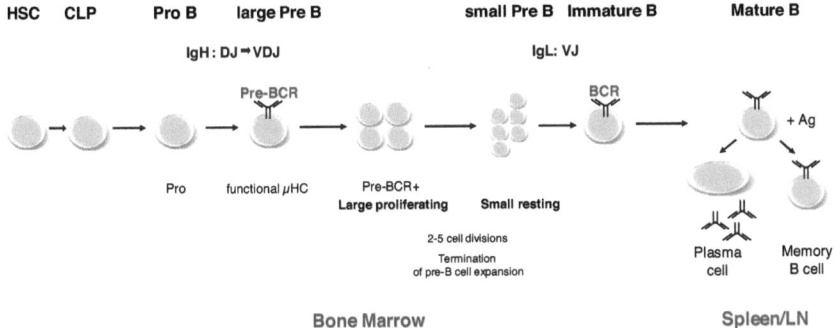

Figure 1. Schematic view of B cell development.
B cells originate in the bone marrow from hematopoietic stem cells. D_H, J_H, V_H are arranged at the pro-B cell stage and functional rearrangement lead to pre-BCR assembly and a burst of proliferation. V_L and J_L are rearranged at the small pre-B cell stage. After assembly of a functional, non autoreactive BCR immature B cells leave the bone marrow. In the secondary lymphoid organs mature B cells can be activated by antigen. They differentiate either into antibody secreting plasma cells (PCs) or memory B cells. HSC: hematopoietic stem cell, CLP: common lymphoid progenitor, Pro: progenitor, Pre: precursor, IgH: heavy chain, IgL: light chain, BCR: B cell receptor, Ag: antigen.

1.1.2 Peripheral B cell development

Out of 2×10^7 cells that develop daily in the bone marrow of the mouse, 10-20% leave the bone marrow (Osmond, 1991; Osmond, 1993), 10% reach the spleen and only 1-3% enter the mature B cell pool (Allman et al., 1993; Melchers et al., 1995).

In the spleen immature B cells are divided into three transitional (T) stages by expression of surface molecules which are summarized in table 1 (Allman et al., 2001; Matthias and Rolink, 2005; Tussiwand et al., 2009).

Table 1. Summary of transitional B cell surface marker expression

	CD19	CD93	IgM	CD21	CD23	IgD
T1	++	+++	+++	-	-	+
T2	++	++	+++	++	++	+++
T3	++	+	++	++	++	+++

Transitional cells are thought to be the direct precursors of the mature naïve B cell compartment (Tussiwand et al., 2009). Transfer experiments into RAG-deficient mice by Loder et al. (Loder et al., 1999) showed that T1 cells can give rise to T2 and mature B cells whereas T2 cells only develop into mature B cells. It is likely that the development occurs step-wise from T1 over T2 into mature B cells but it is not excluded that T1 B cells can also directly develop into mature B cells (Allman and Pillai, 2008). Whether T3 B cells are an developmental intermediate or are anergic B cells as suggested by Merrel et al. (Merrell et al., 2006) is still under debate (Allman and Pillai, 2008). From T2 or T3 B cells, cells develop either into follicular (Fo) or marginal zone (MZ) B cells.

For the cell fate decision between Fo and MZ B cells several signalling events are essential. Firstly, BCR specificity plays a key role since immature B cells specific for phosphorylcholine, an antigen of encapsulated bacteria compromising a significant fraction of the gut flora, selectively yield MZ B cells (Martin and Kearney, 2000). Secondly, BCR signal strength is important. Mutations resulting in diminished BCR signalling can leave the MZ pool intact wheras B1 and Fo B cells are lost (Cariappa et al., 2001). Thirdly, Notch2 signalling via Delta like-1 (DL-1) a Notch2 ligand is essential for MZ B cell differentiation (Hozumi et al.,

2004; Saito et al., 2003). It is furthermore proposed that BCR, BAFF-R and Notch signalling collaborate to promote B cell activation and MZ B cell development by amplifying and/or sustaining NF-κB activation, as p50+/- and Notch2+/- mice show a complete loss of MZ but not Fo B cells (Moran et al., 2007). Conversely, MINT (MSH-homeobox-homologue 2-interacting nuclear target) expression preventing Notch2 signalling (Kuroda et al., 2003) and Bruton's tyrosine kinase (BTK) expression (Loder et al., 1999; Makowska et al., 1999) lead to preferential development of Fo B cells. A very recent study also suggests Foxo1 as a player in the choice between Fo and Mz B cells since Foxo1 B cell-specific deletion leads to an increased MZ B cell compartment (Chen et al., 2010).

90% of mature B cells are follicular B cells which circulate through lymph and blood stream occupying lymph nodes, peyers patches and the spleen. For entry in peripheral and mesenteric lymph nodes as well as peyers patches through high endothelial venules (HEVs), $\alpha_4\beta_7$ integrin (Moran et al., 2007), L-selectin (also called CD62L) and CCR7 are required (Okada et al., 2002). In contrast to HEVs, B cells entering the spleen are passively released in sinuses of the marginal zone and red pulp (Stein and Nombela-Arrieta, 2005).

Fo cells build the major part in thymus-dependent (TD) immune responses (Allman and Pillai, 2008). In the spleen they are positioned in the follicles via expression of CxCR5 (Ansel et al., 2000). Early in a T cell dependent response Fo B cells can either differentiate into short-lived plasma cells or enter germinal centers (GCs). Here, they differentiate into plasmablasts, exhibit somatic hypermutation and class switch recombination where activation induced deaminase (AID) is indispensable (Honjo et al., 2002; Muramatsu et al., 2000). Emerging plasma cells (PCs) gain the ability to migrate to the bone marrow to become long-lived plasma cells by interaction of APRIL and/or BAFF produced by stromal cells with BCMA, a BAFF-receptor, on plasma cells (O'Connor et al., 2004).

In contrast, MZ B cells reside in the spleen at the marginal sinuses between white and red pulp where they are located by expression of S1P1 and S1P3 (Cinamon et al., 2004; Cinamon et al., 2008; Vora et al., 2005). MZ B cells encounter preferentially thymus-independent (TI) blood borne antigens via high expression of CD21 and can migrate into the red pulp to differentiate into short lived plasma cells (Martin and Kearney, 2002). In addition, they participate in T-dependent immune responses by transporting TD antigens into follicles where they can activate Fo B and T helper cells (Attanavanich and Kearney, 2004; Cinamon et al., 2008; Song and Cerny, 2003). This migration is accompanied by S1P1 and S1P3 downregulation which leads to attraction of MZ B cells via CxCR5 into the follicles (Cinamon et al., 2004). High expression of CD1d is thought to be involved in lipid antigen presentation to NKT cells which in return may activate MZ B cells via CD40L-CD40 interactions (Leadbetter et al., 2008).

A third peripheral B cell population are B1 cells. B1 cells preferentially reside in the peritoneal and pleural cavity. They are further distinguished into B1a and B1b cells which can be separated by the surface expression of CD5 (B1a cells are $CD5^+$ and B1b cells are $CD5^-$). B1a B cells are thought to develop from fetal liver precursors with a restricted BCR repertoire (Dorshkind and Montecino-Rodriguez, 2007) whereas B1b cells may originate from B2 cells in a T-independent manner and represent a specialized type of IgM memory cells (Alugupalli et al., 2004). B1 B cells can contribute to the generation of IgM responses to TI antigens such as phosphorylcholine, an antigen on many pathogenic bacteria (Allman and Pillai, 2008; Alugupalli et al., 2003; Boes et al., 1998) and have long thought to be the main producer of natural serum IgM which have a low affinity and include autoreactive species (Baumgarth et al., 2000; Briles et al., 1982; Ochsenbein et al., 1999; Su et al., 1991). However, a recent study proposed that the natural serum IgM is produced mostly by splenic B1 B cells in an IRF4-dependent but T-independent manner (Holodick et al., 2010).

B1 cells can migrate from the peritoneum to mesenteric lymph nodes and to the intestinal lamina propria. One proposed mechanism includes the migration via the omentum, parathymic lymph nodes through lymphatics and the thoracic duct into the blood stream (Ansel et al., 2002) whereas another mechanism implies direct trafficking to the intestine via the omentum. However, the contribution of S1P1 in mediating B1 egress is presently unclear (Ha et al., 2006; Kunisawa et al., 2007). At mucosal sites B1 B cells can contribute to the generation of T-independent IgA responses (Allman and Pillai, 2008). Furthermore, B1 cells have self renewal potential but whether traffic through the spleen is required for self-renewal remains controversial (Ha et al., 2006; Kunisawa et al., 2007).

1.1.3 Plasma cell generation and migration

Plasma cells are the only type of cells which produce antigen-specific antibodies. After antigen encounter plasma cells can develop from naïve marginal zone, follicular B cells or activated germinal center B cells and memory B cells (Fig. 2). Whether MZ, Fo or GC B cells differentiate into PCs depends on the origin, dose and form of the antigen as well as on location of antigen encounter (Shapiro-Shelef and Calame, 2005). Marginal zone B cells preferentially recognize blood borne T-independent antigens type 2 (TI-2). Within a few hours after immunization with a TI-2 antigen, MZ B cells move to the red pulp for proliferation and plasmablast differentiation (Lopes-Carvalho and Kearney, 2004). They react more rapidly and with a lower BCR threshold than follicular B cells (Oliver et al., 1997) presumably due to expression of CD21, CD1d, CD38, CD80 and CD86 (Oliver et al., 1997; Oliver et al., 1999). In contrast, Fo B cells preferentially react to T-dependent protein antigens. Two days after immunization foci of plasmablasts are observed along the periphery of periarteriolar lymphoid sheath which expand until 8 days after immunization and then disappear (Jacob et al., 1991). These short-lived plasma cells building the rapid initial immune response develop either from MZ or Fo B cells without somatic hypermutated Immunoglobulins and express the plasma cell differentiation factors IRF4, Blimp-1 and XBP1 (Mittrucker et al., 1997; Reimold et al., 2001; Shapiro-Shelef et al., 2003).

After antigen encounter and receiption of T cell help, some Fo B cells migrate into the B cell follicles and form germinal centers (McHeyzer-Williams et al., 2001; McHeyzer-Williams, 2003). In GCs activated Fo B cells undergo several rounds of proliferation which is accompanied by affinity maturation and class switch recombination (CSR). Cells are kept in the GC state by expression of BCL6 (Cattoretti et al., 1995; Dent et al., 1997; Fukuda et al., 1997) and Bach2 (Muto et al., 2004) which repress Blimp-1 until germinal center reaction is completed (Shaffer et al., 2000; Tunyaplin et al., 2004). PCs exit diminishing GCs 10 to 14 days later by downregulation of CxCR5 and CCR7. GC-derived PCs express Blimp-1, XBP1 and IRF-4 (Shapiro-Shelef and Calame, 2005), present somatically mutated, high affinity BCRs and express switched immunoglobulin isotypes (Hargreaves et al., 2001). Emerging memory B cells do not secrete antibodies and can persist independently of antigenic stimulation (Maruyama et al., 2000; McHeyzer-Williams and Ahmed, 1999). These cells can rapidly respond to a second antigen encounter.

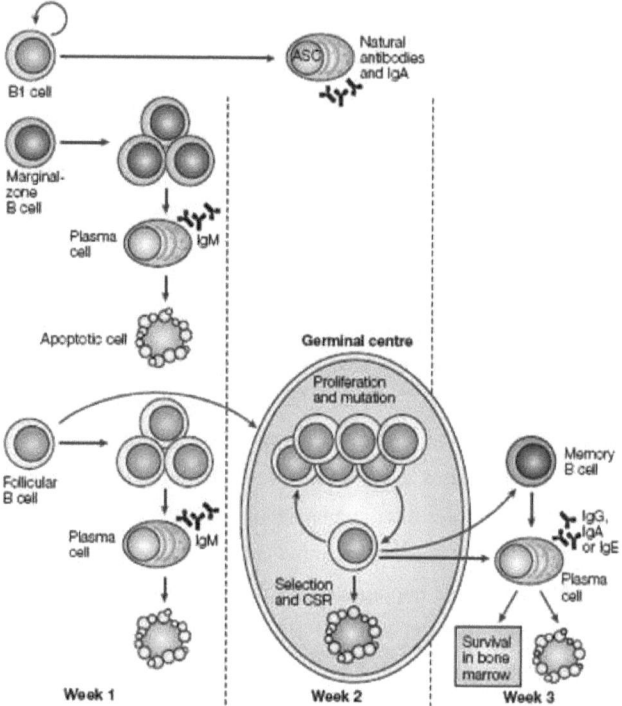

Figure 2. Formation of plasma cells.
B1 cells secrete natural antibodies in the absence of external antigen and can give rise to IgA antibody-secreting cells (ASCs) in the gut upon pathogen encounter. On foreign antigen encounter, naïve MZ B and Fo B cells differentiate into plasma cells. Most of the extrafollicular plasma cells that are formed in this early response are short-lived. Some activated follicular B cells form a germinal center. Plasma cells that emerge from a germinal-centre reaction might become long-lived if they find survival niches, which are mainly located in the bone marrow or develop into memory B cells (adapted from Shapiro-Shelef and Calame, 2005).

PCs which have undergone GC reactions build the pool of long-lived plasma cells. These cells migrate either to the bone marrow, the mucosa or sites of inflammation (Kunkel and Butcher, 2003). For splenic egress PCs need to upregulate S1P1 (Kabashima et al., 2006). For migration to the bone marrow PCs upregulate the chemokine receptors CxCR4 and presumably CxCR3 as well as the integrin $\alpha_4\beta_1$ (Hargreaves et al., 2001; Hauser et al., 2002; Kunkel and Butcher, 2003; Odendahl et al., 2005). $\alpha_4\beta_1$ intergin binds vascular cell adhesion molecule1 (VCAM1) at the cell surface of bone-marrow epithelial cells. Once having reached the bone marrow PCs are held in their survival niches by IL-6 (Minges Wols et al., 2002), BAFF/Blys (O'Connor et al., 2004), IL-5, CxCL12 (Nie et al., 2004), TNF and CD44 ligands (Cassese et al., 2003).

IgA secreting PCs addressed for the mucosa express CCR9 (small intestine), CCR10 (all mucosal compartments) (Mora and von Andrian, 2008) and $\alpha_4\beta_7$ integrin which binds mucosal vascular addressin cell-adhesion molecule1 (MAdCAM1) presented on intestinal epithelial cells (Kunkel and Butcher, 2003). However, the exact mechanisms underlying PC migration and survival are still under extensive investigation.

1.2 Krüppel-like factor 2 (KLF2)

1.2.1 Krüppel-like transcription factors

B cell development is strictly regulated by expression and repression of transcription factors, adhesion molecules as well as signalling components. We recently identified KLF2 as a target gene of pre-BCR signaling during early B cell development (Schuh et al., 2008). KLF2/LKLF belongs to the family of krüppel like transcription factors, which consists of at least 17 members in mammals and is closely related to drosophila Krüppel protein (Pearson et al., 2008). Krüppel-like transcription factors bind to GC rich DNA domains via three c-terminal zinc fingers and are involved in controlling proliferation and terminal differentiation of various cell types (Pearson et al., 2008). KLF2 was originally discovered in lung tissue and was shown to be important for heart function (Kaczynski et al., 2003). Apart from its role in lung and heart development, KLF2 plays an important role during development and activation of T lymphocytes (Kuo et al., 1997) and monocytes (Das et al., 2006).

1.2.2 KLF2 in T lymphocytes

KLF2 is expressed in naïve T cells, downregulated upon activation and re-expressed in memory T cells (Bai et al., 2007; Grayson et al., 2001; Schober et al., 1999; Wu and Lingrel, 2005), thereby regulating the expression of c-myc and p21 (Buckley et al., 2001; Haaland et al., 2005; Wu and Lingrel, 2004) (Fig. 3). In addition, overexpression of KLF2 results in a c-myc dependent block of cell cycle progression in a transformed T cell line (Buckley et al., 2001).

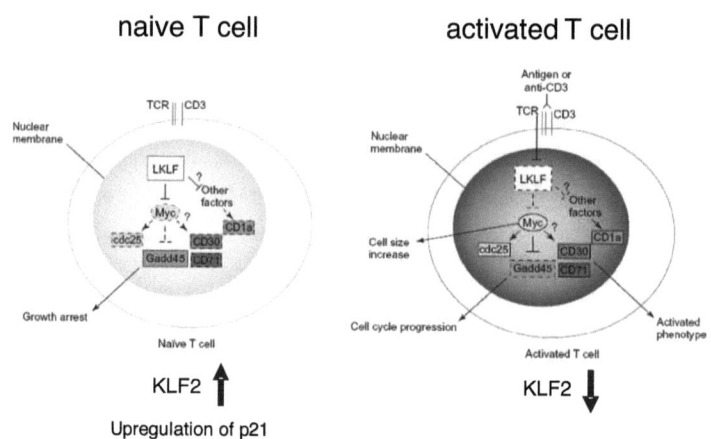

Figure 3. Role of KLF2 in programming T cell quiescence.
KLF2 promotes a quiescent phenotype in T cells by downregulating the expression of *myc* and upregulation of p21. C-mMyc promotes cell cycle progression by upregulating genes, such as cdc25, which encodes a phosphatase shown to activate cyclin-dependent kinase-2 (cdk-2). C-myc also downregulates the expression of Gadd45 (growth arrest and DNA damage-induced gene 45), which has growth inhibitory effects. Expression of c-myc in LKLF-deficient cells increases the levels of CD30 and CD71, but not CD1a, suggesting that LKLF promotes quiescence through other factors, such as p21 in addition to c-myc (adapted from Yusuf and Fruman, 2003). Lines indicate function or expression. Broken lines indicate loss of function or expression. TCR: T-cell receptor.

Furthermore, exogenous KLF2 expression in activated T cells promotes cell migration to blood and lymph nodes (Bai et al., 2007), whereas KLF2-deficient T cells accumulate in the thymus due to an emigration defect consistent with decreased levels of S1P1 (Carlson et al., 2006; Kuo et al., 1997). In a previous study it was shown that KLF2 binds directly to the S1P1 and L-selectin promotor and regulates their expression in T cells (Bai et al., 2007; Carlson et al., 2006). Additionally, KLF2 upregulates $\alpha_4\beta_7$-integrin in T cells (Carlson et al., 2006). The role of KLF2 in regulating chemokine receptor expression in T cells is controversially discussed in two recently published studies. *Vav-cre* and *lck-cre* mediated deletion of KLF2 in T cells resulted in upregulation of CCR3 and CCR5 (Sebzda et al., 2008), whereas *CD4-cre* mediated deletion led to upregulation of CxCR3 (Weinreich et al., 2009). However it was clearly shown that KLF2 controls cytokine production (IL4) in T cells (Weinreich et al., 2009).

Taken together, KLF2 is an important regulator of T cell migration, homing and activation.

1.2.3 Regulation of KLF2

The mechanisms regulating KLF2 activitiy in B cells remain unclear. In a previous study we identified KLF2 as a late target gene of pre-BCR signalling during early B cell development (Schuh et al., 2008). Microarray studies revealed that KLF2 transcripts are highly abundant in resting B cells, downregulated upon mitogenic activation and re-expressed in plasma and memory B cells (Bhattacharya et al., 2007; Glynne et al., 2000; Kabashima et al., 2006).

One recent study showed that Foxo1 can bind the KLF2 promotor and activate transcription (Fabre et al., 2008). In addition, Gubbels Bupp et al. showed reduced KLF2 transcripts after conditional deletion of Foxo1 in T cells (Gubbels Bupp et al., 2009). Foxo1 is important at many stages of B cell development. Foxo1 induces RAG expression (Su et al., 2003) and gets downregulated upon pre-BCR induction via phosphorylation by downstream signalling of Phosphoinositide 3-kinase (PI3K). Furthermore, Foxo1 is essential for mature B cell survival, trafficking and class switch recombination (Dengler et al., 2008; Omori et al., 2006).

Phosphoinositide 3-kinases (PI3Ks) are an evolutionarily conserved family of lipid-modifying enzymes which can be divided in three classes (class I, II and III) (Deane and Fruman, 2004). Class IA members are activated upon phosphorylation cascades following pre-BCR and BCR signalling including Src- and Syk-family proteins, CD19 and BCAP (B-cell PI3K adaptor protein) (reviewed in Werner et al., 2010). Activated PI3Ks consist of a regulatory and a catalytic subunit whereas the combination of p85α and p100δ respectively are thought to be the main class in pre-BCR and BCR signalling (Clayton et al., 2002; Jou et al., 2002; Okkenhaug et al., 2002; Suzuki et al., 1999). An initial event upon PI3K activation is generation of the lipid second messenger phosphatidylinositol-(3,4,5)-phosphate (PtdInsP3) (Okkenhaug and Vanhaesebroeck, 2003). PtdInsP3 recruits proteins like the serine/threonine protein kinase B (PKB or Akt) or BTK to the membrane and leads to their subsequent activation (Werner et al., 2010). Whereas BTK is important for calcium signalling via activation of phospholipase Cγ (PLCγ), Akt is involved in regulation of cell cycle, survival and differentiation (Herzog et al., 2009; Manning and Cantley, 2007). For example, Akt phosphorylates Foxo proteins (Foxo1, 3a and 4) which leads to nuclear export and proteosomal degradation (Brownawell et al., 2001; Burgering and Kops, 2002; Rena et al., 2001) or inactivates the glycogen synthase kinase-3β, a negative regulator of c-myc and cyclin D leading to cell cycle entry (Vivanco and Sawyers, 2002).

Furthermore, the group of Jerry B. Lingrel could show an interaction of KLF2 with the E3 ubiquitin ligase WWP1 (Conkright et al., 2001; Zhang et al., 2004) which binds to the autoinhibitory domain of KLF2 in COS-1 cells and leads to proteasomal degradation by ubiquitiniation of Lysine 121 (Zhang et al., 2004). WWP1 belongs to the HECT domain E3 ligases (Hatakeyama and Nakayama, 2003) and its C. elegans homolog appears to play a role in embryogenesis (Huang et al., 2000) and transcriptional regulation (Mosser et al., 1998).

2 Results

2.1 B cell speficic KLF2 deletion *in vivo*

2.1.1 KLF2 expression profile during B cell development

Within the B cell lineage, KLF2 is expressed in small resting pre-B cells, naïve B cells and plasma as well as memory B cells (Bhattacharya et al., 2007; Glynne et al., 2000; Kabashima et al., 2006; Schuh et al., 2008). Since these findings were mainly based on microarray data, we analyzed KLF2 expression in B cell lines, resting and activated B cells on mRNA and protein level. As shown in Fig. 4A KLF2 transcripts could be detected in all tested B cell lines (Fig. 4A) whereas KLF2 protein was only present in WEHi 279, WEHi 231 (CP) and NYC cells (Fig. 4B). S1P1 transcripts were found in all cell lines were the protein was detected and additionally in the pro-B cell line 38B9, whereas Tk- and Tkμ cells made only KLF2 transcripts but not S1P1 transcripts, suggesting that S1P1 expression is not strictly dependent on KLF2 in B cells (Figs. 4A, B).

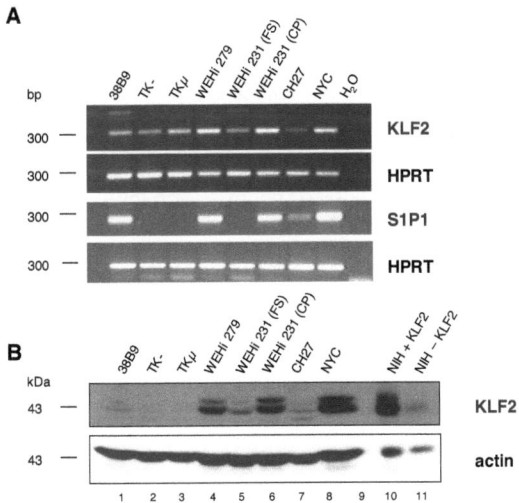

Figure 4. Analyses of KLF2 expression profile in different B cell lines.
(A) RT-PCR analyses of total RNA from different B cell lines with gene-specific primers for KLF2 and S1P1. HPRT signals control the integrity and loading of cDNA. **(B)** Western-Blot analyses of KLF2 expression in different B cell lines and NIH+/-KLF2. Membranes were stained with an anti-KLF2 antiserum and an anti-actin antiserum as loading control.

Freshly isolated CD43 negative B cells displayed high amounts of KLF2 transcripts, which are drastically decreased upon treatment with either LPS or anti-CD40/IL4/anti-IgM (αBCR) and upregulated again in sorted bone marrow plasma cells (Fig. 5A). In parallel, high amounts of KLF2 protein were detected in freshly isolated CD43 negative B cells. Two forms of KLF2 protein were visible in resting B cells, suggesting that KLF2 protein is posttranslationally modified. Upon activation with either LPS or αBCR treatment, we found a rapid decrease in the amount of KLF2 protein (Figs. 5B, C) with the higher molecular weight form of KLF2 disappearing faster. This was already seen in the short time treatment with LPS (Fig. 5B) and much clearer upon 48h αBCR stimulation where the higher molecular weight form was completely absent (Fig. 5C). More surprisingly, although abundant in follicular (Fo) B cells, KLF2 protein is barely detectable in purified marginal zone (MZ) B cells (Fig. 5D).

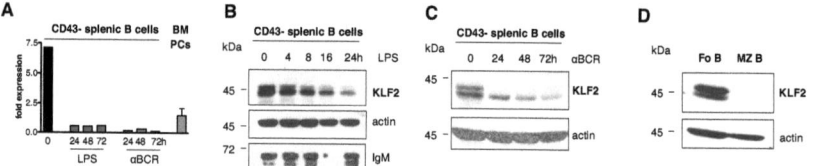

Figure 5. Analyses of KLF2 expression profile in B cells.
(A) Quantitative TaqMan PCR of total RNA from CD43 depleted sorted splenic B cells upon mitogenic activation with either LPS (10 μg/ml) or anti-CD40/IL4/anti-IgM (αBCR: 10 μg/ml, 100 U/ml, 10 μg/ml) at different time points (0, 24, 48, 72 h) and bone marrow plasma cells with a gene-specific TaqMan probe for KLF2. A TaqMan probe for HPRT was used to calculate fold expression. (B-D) Western-Blot analyses of KLF2 expression in purified splenic B cells (CD43 depletion) upon mitogen activation with either LPS (B) or αBCR (C) at different time points and in (D) sorted Fo and MZ B cells. Membranes were stained with an anti-KLF2 antiserum, an anti-IgM serum as stimulation control (B) and an anti-actin antiserum as loading control.

In summary, we showed that KLF2 is upregulated upon pre-BCR stimulation, downregulated after activation and is rexpressed in PCs (Fig. 6).

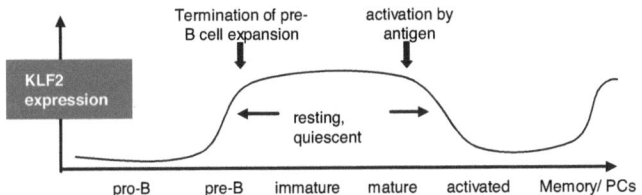

Figure 6. Schematic view of KLF2 expression profile during B cell development.
KLF2 is expressed upon pre-BCR induction and stays expressed during further maturation to mature B cells. After activation with thymus-independent and –dependent antigens KLF2 is downregulated and reexpressed in memory and plasma cells (PC).

2.1.2 Normal B cell development in the bone marrow of KLF2-deficient animals

One of the presumed functions of KLF2 is a role in maintaining the quiescent state of a cell since ectopic overexpression of KLF2 in a transformed T cell line blocked cell cycle progression by modulating the expression of c-myc and p21 (Buckley et al., 2001; Haaland et al., 2005; Wu and Lingrel, 2004). Applied to B lymphoid cells, an increase of KLF2 expression upon pre-BCR induction would terminate the cell cycle of proliferating pre-B cells (Schuh et al., 2008; Vettermann and Jäck, 2010). Accordingly, KLF2 deficiency should result in hyperproliferation of pre-B cells, and thus, in an increase of the pre-B cell pool. To test this hypothesis, we established B cell-specific KLF2 knockout mice by crossing mice carrying a floxed KLF2 ($KLF2^{flox}$) allele (Lee et al., 2006) (Fig. 7A) with the B cell-specific deleter strain *mb1-cre* (Hobeika et al., 2006). Genotypes were tested by PCR for wt, floxed and mb1 cre alleles (Fig. 7B). In these mice, deletion of KLF2 occurs in early B cell precursors, as demonstrated by western blot analyses of CD19 positive B lymphoid cells from bone marrow and spleen (Fig. 7C). Since *floxed* (fl) mice already showed a reduced synthesis of KLF2 which was comparable with heterozygous (wt/-) mice, wildtype mice +/- *mb1 cre* were used as littermate controls (Figs. 7D, 8).

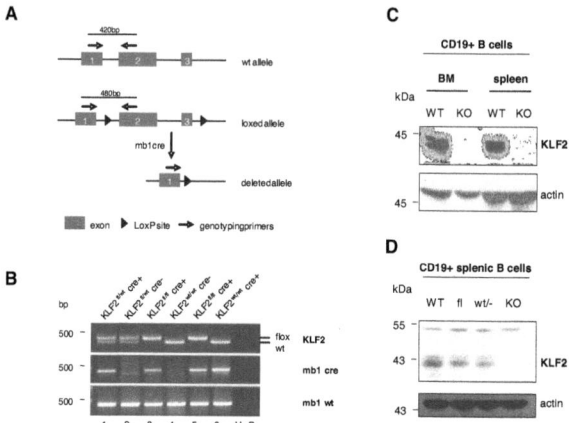

Figure 7. KLF2 conditional knockout strategy.
(A) Schematic view of the wt and the floxed KLF2 allele with exon 2 and exon 3 flanked by LoxP sites (Lee et al., 2006). Sizes of PCR amplification products are indicated. **(B)** PCR analysis of total tail DNA with gene-specific primers for $KLF2^{fl/wt}$, mb1 cre and mb1 wt. **(C, D)** Western-Blot analyses of CD19-positive sorted bone marrow and splenic B cells from wildtype (WT) and KLF2-deficient (KO) mice **(C)** and sorted CD19 positive splenic B cells from different genotypes ($KLF2^{wt/wt}$ as WT, $KLF2^{wt/-}$ as wt/-, $KLF2^{flox/flox}$ as fl and $KLF2^{-/-}$ as KO) **(D)**. Membranes were stained with an anti-KLF2 antiserum and an anti-actin antiserum as loading control.

In flow cytometric analyses, the frequencies and numbers of pro-B ($CD19^+$/c-kit$^+$), pre-B ($CD19^+$/$CD25^+$) and immature $B220^+$/IgM^+ B cells (Fig. 8) did not differ significantly between KLF2-deficient and -sufficient mice. However, frequencies and numbers of recirculating mature B cells ($B220^{hi}$/IgM^+) are decreased by half, indicating that KLF2-deficient mature B cells might have a defect in their homing capacity to the bone marrow.

In summary, KLF2-deficient animals display normal B cell development in the bone marrow. However, numbers and frequencies of recirculating B cells were decreased.

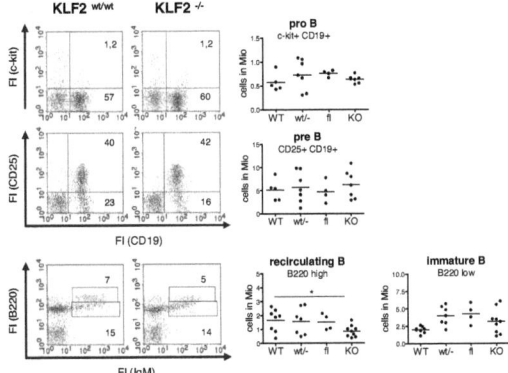

Figure 8. Analysis of B cell precursors and recirculating B cells in the bone marrow from wildtype and KLF2-deficient mice.
Flow cytometric analyses of pro- (CD19$^+$/c-kit$^+$), pre- (CD19$^+$/CD25$^+$), recirculating (B220^{hi}/IgM$^+$) and immature (B220^{low}/IgM$^+$) B cells. FACS dot plots show one representative experiment and scatter plots implement all mice measured and mean values from all genotypes (KLF2$^{wt/wt}$ as WT, KLF2$^{wt/-}$ as wt/-, KLF2$^{flox/flox}$ as fl and KLF2$^{-/-}$ as KO).

2.1.3 Enlarged spleen size and increased numbers of splenic B cell populations in KLF2-deficient animals

Immature B cells leave the bone marrow and enter the blood stream. Once they have entered the spleen, they develop into transitional B cells (Allman et al., 1992) and differentiate either into mature Fo or MZ B cells (Martin and Kearney, 2002). Fo B cells, which are also present in other lymphoid tissues, react to TD antigen and can leave the spleen. In contrast, MZ B cells remain in the marginal zone of the spleen reacting mainly in response to TI blood-borne antigens (Cyster, 2000). When we compared B cell subsets in the spleen from KLF2-deficient and wildtype mice, we observed profound effects of KLF2 deletion on peripheral B cell homeostasis (Figs. 9, 10). KLF2-deficient mice have enlarged spleens with a 2-3fold increase in total cell number but identical mouse weights (Fig. 9).

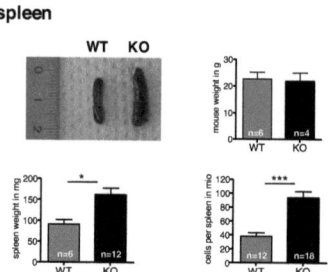

Figure 9. Analysis of of spleen physiognomy from wildtype and KLF2-deficient mice.
Comparison of wildtype and KLF2-deficient spleens: mice and spleen weight and cell numbers in mio of total spleens are shown.

In flow cytometric analyses we found a 2-3fold increase in the total number of B220$^+$/IgM$^+$ splenic B cells, an increase in all transitional B cell stages (T1-T3), a 2-3fold increase in numbers of Fo B (B220$^+$/CD21^{lo}/CD23^{hi}), a 4-5fold increase of MZ B cells (B220$^+$/CD21^{hi}/CD23$^{lo/neg}$) and a 2fold increase of B1 cells (CD5$^+$/IgM$^+$) (Fig. 10).

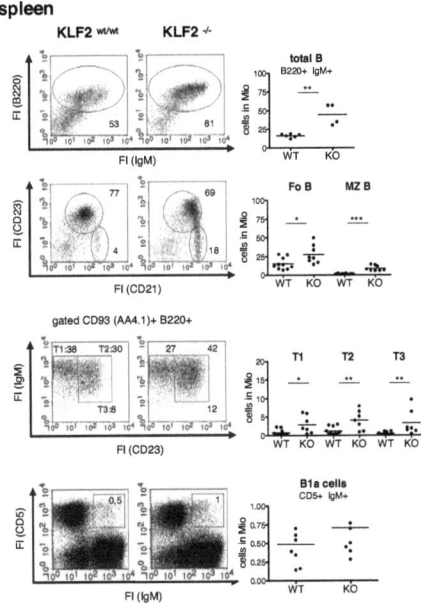

Figure 10. Analysis of B cells in spleen from wildtype and KLF2-deficient mice.
Flow cytometric analyses of total (B220$^+$/IgM$^+$), Fo (CD23$^+$/CD21^{low}), MZ (CD23$^-$/CD21^{hi}), transitional B (T1: IgM^{hi}/CD23$^-$, T2 IgM^{hi}/CD23$^+$, T3 IgM^{low}/CD23$^+$) and B1 cells (CD5$^+$/IgM$^+$). FACS dot plots show one representative experiment and scatter plots implement all mice measured and mean values.

This increase is not due to increased survival and/or proliferation, since freshly isolated KLF2-deficient Fo and MZ B cells showed the same survival capacity as the corresponding wildtype B cell subsets (Fig. 11A); in addition, frequencies of 5-bromo-2-deoxyuridine (BrdU)-positive pro-B, pre-B in the bone marrow as well as Fo B, and MZ B cells in the spleen did not differ between wildtype and KLF2-deficient mice upon a single intraperitoneal injection of BrdU (Fig. 11B).

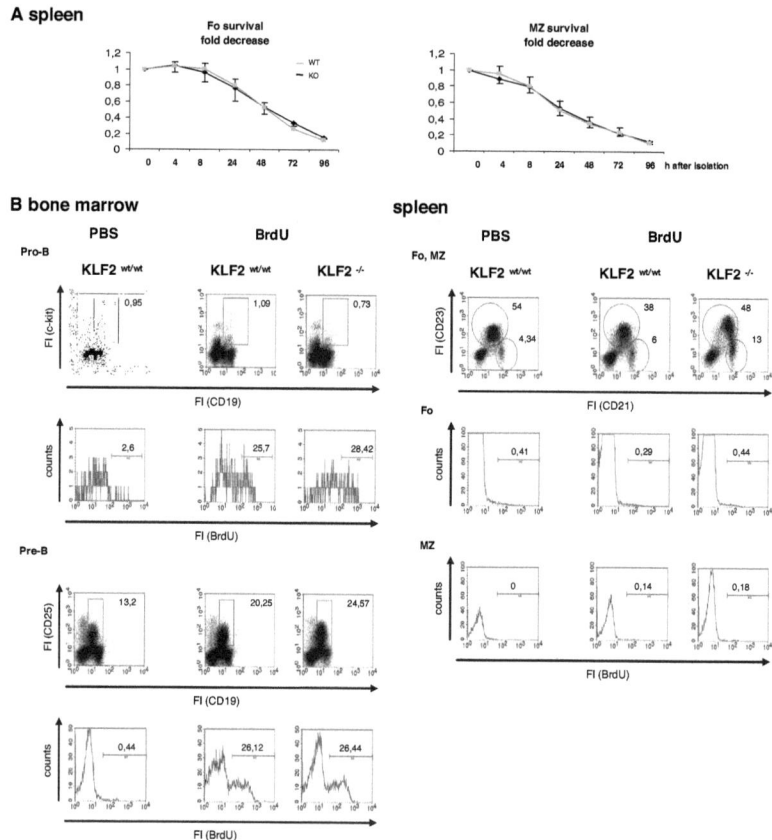

Figure 11. KLF2 deficiency does not affect *in vitro* survival capacity and proliferation of Fo and MZ B cells.
(A) Freshly isolated Fo and MZ B cells from WT and KLF2-deficient animals were cultured in duplicates in RPMI supplemented with 10% FCS over a period of 96 hours. The number of living cells was determined by flow cytometry. Graphs show mean fold decrease of Fo and MZ B cell numbers from 6 wildtype and 4 KLF2-deficient animals. **(B)** Mice were injected with 2 mg BrdU intraperitoneally and analyzed for BrdU incorporation 20 h later by flow cytometry. Shown is one representative dotplot for the B cell population analyzed and the corresponding histogram showing BrdU incorporation of gated B cell populations: pro-B (c-kit$^+$/CD19$^+$) and pre-B (CD25$^+$/CD19$^+$) from bone marrow and Fo B (B220$^+$/CD23$^+$/CD21^{low}) and MZ B (B220$^+$/CD23$^-$/CD21^{hi}) from spleen. One representative dotplot is shown of a PBS injected wildtype control to the left (n=1), a wildtype mouse in the middle (n=4) and a KLF2-deficient mouse to the right (n=2).

Since we found increased numbers of Fo and MZ B cells in the spleen of KLF2-deficient mice, we examined the architecture of splenic follicles on cryosections. B cell follicles in the spleen of wildtype mice consist of a core with IgD^{hi}/IgM^{lo} Fo B cells, separated by a ring of MOMA$^+$ metallophilic macrophages from the marginal zone harboring IgD^{lo}/IgM^{hi} MZ B cells (Fig. 12). In the spleen of KLF2-deficient mice, B cell follicles with Fo and MZ B cells are present; but they are larger in size, and the separation between MZ and Fo B cells is not as clear as in KLF2-deficient mice (Figs. 12B, C).

In summary, KLF2 seems to be important for controlling the separation of Fo and MZ B cells within splenic compartments.

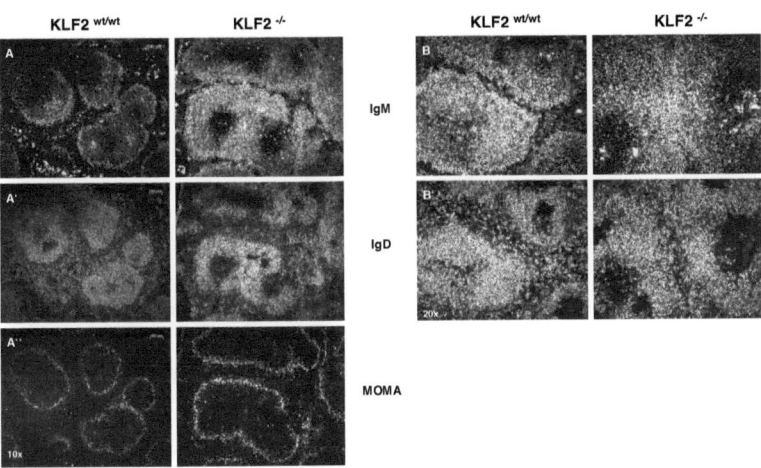

Figure 12. Analysis of spleen architecture from wildtype and KLF2-deficient mice.
(A-C) Acetone fixed tissue cryosections were stained with the indicated fluorochrome conjugated antibodies (stained surface markers are indicated in the middle).

2.1.4 Decreased frequencies of B cells in the peritoneal cavity, the blood and peyers patches of KLF2-deficient mice

Mature Fo B cells not only home to the spleen but circulate through the body in the blood- and lymphstream and enter other peripheral lymphatic organs (Allman and Pillai, 2008). To determine whether KLF2 controls migration and homing of mature peripheral B cell subsets, we analyzed inguinal lymph nodes, liver, thymus, blood, peyers patches and the peritoneal cavity for the presence of B cells. By flow cytometric analyses of inguinal lymph nodes and liver, we found no differences in numbers and frequencies of CD19$^+$/IgM$^+$ B cells (Figs. 13A, B).

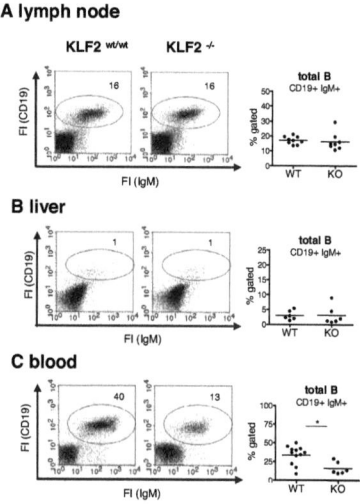

Figure 13. Analysis of B cells in lymph node, liver and blood from wildtype and KLF2-deficient mice.
(A-C) Flow cytometric analyses of total (CD19$^+$/IgM$^+$) B cells in the inguinal lymph node **(A)**, the liver **(B)**, and blood **(C)**. FACS dot plots show one representative experiment and scatter plots implement all mice measured and mean values.

In addition, T and B cell populations were unaffected in the thymi of KLF2-deficient mice in comparison to littermate controls as shown by flow cytometric analyses (Fig. 14).

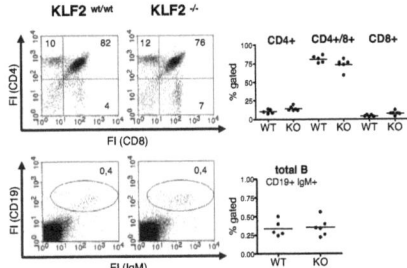

Figure 14. Analysis of thymi from wildtype and KLF2-deficient mice.
Flow cytometric analysis of T cells (CD4$^+$, CD4/CD8$^+$ and CD8$^+$, upper panel) and total B cells (CD19$^+$/IgM$^+$ lower panel) in the thymus. FACS dot plots show one representative experiment and scatter plots implement all mice measured and mean values.

In contrast, frequencies of B cells in the blood of KLF2-deficient animals are decreased 2-3fold (Fig. 13C). Similarily, peyers patches in KLF2-deficient animals are smaller, their numbers are reduced (Fig. 15A) and they contain 2-3 times less CD19$^+$/IgM$^+$ B cells (Fig. 15B). Consistent with the reduced number of peyers patches and B cells, serum level of "natural" IgA are reduced, as determined by ELISA (Fig. 17).

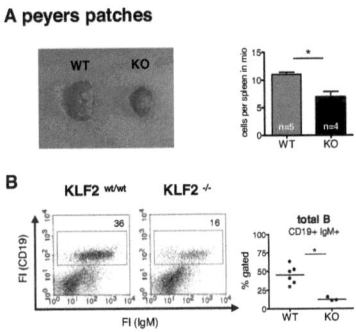

Figure 15. Analysis of peyers patches from wildtype and KLF2-deficient mice.
(A) Analysis of morphology and number of peyers patches. (B) Flow cytometric analyses of total (CD19$^+$/IgM$^+$) B cells in peyers patches. FACS dot plots show one representative experiment and scatter plots implement all mice measured and mean values.

The effect of KLF2 deletion on B1 cells in the peritoneal cavity is even more severe, with a clear reduction in B1a (IgM+/CD5+) and B1b (IgM+/CD5-/CD11b+) cells (Fig. 16). Since B1 cells are the major source for production of "natural" IgM in the serum (Berland and Wortis, 2002), we determined the amount of serum Ig titers by ELISA, but could not detect any differences in "natural" IgM or IgG serum levels (Fig. 17).

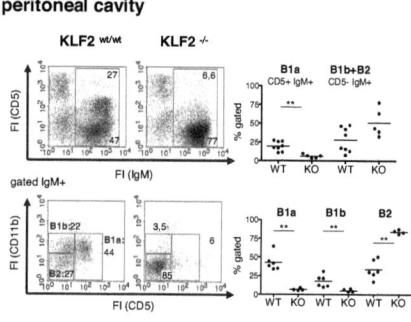

Figure 16. Analysis of peritoneal B cell subsets from wildtype and KLF2-deficient mice.
Flow cytometric analyses of B1a (CD5+/IgM+), B1b+B2 (CD5-/IgM+), B1b (CD5-/CD11b+) and B2 (CD5-/CD11b-) cells. FACS dot plots show one representative experiment and scatter plots implement all mice measured and mean values.

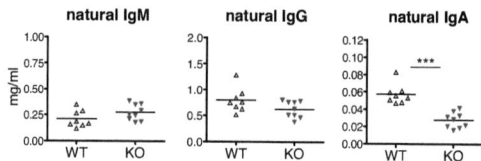

Figure 17. Analysis of antibody titers in non-immunized wildtype and KLF2-deficient mice.
Sandwich-ELISA for different antibody isotypes (IgM, IgG and IgA) of sera from unimmunized wildtype and KLF2-deficient mice. Each dot represents one mouse.

In summary, KLF2 plays a critical role in migration and homing of peripheral B cell subsets. In the absence of KLF2, B1 cells in the peritoneum are almost absent. Moreover, KLF2 deficiency results in a decreased number of peyers patches and low IgA serum titers, suggesting a role of KLF2 in proper assembly of gut-associated lymphoid tissues.

2.1.5 Reduced numbers of plasma cells in the bone marrow after boost immunization with TD antigen

To investigate whether KLF2-deficient B cells can participate in humoral immune responses we immunized wildtype as well as KLF2-deficient animals with either TI or TD antigens (Figs. 18, 19, 20). Several days after immunization, serum samples were collected and Ig titers were analyzed by antigen-specific ELISA. Despite an increase in numbers of Fo and MZ B cells in the spleen, antigen-specific antibody IgM and IgG titers in response to TI-1 (TNP-LPS, Fig. 18A) and TI-2 (TNP-ficoll, Fig. 18B) were fairly normal in KLF2-deficient animals, with the exception of a slight but significant increase in IgG titers against TNP-LPS at day 14 after immunization (Fig. 18A).

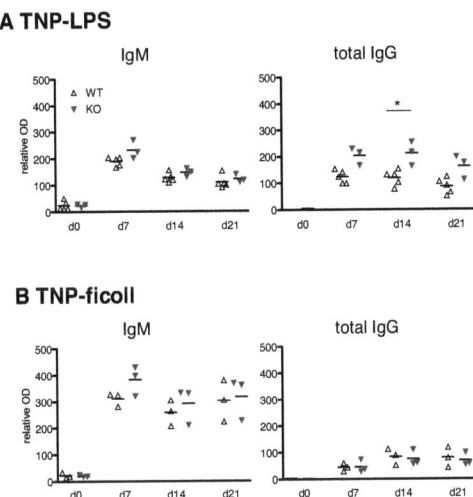

Figure 18. Analyses of Ig-titers after immunization of wildtype and KLF2-deficient mice with T-independent antigens.
Sera of wildtype (blue open triangles) and KLF2-deficient (red filled triangles) mice were collected after i.v. immunization with 50 µg TNP-LPS (A) or 25 µg TNP-ficoll (B) and analyzed by ELISA for antigen specific antibody production (IgM, IgG).

Similarly, IgM and IgG responses against the TD antigen TNP-coupled keyhole limbet hemocyanin (TNP-KLH) did not differ significantly between wildtype and KLF2-deficient mice

14 days after primary immunization (Fig. 19A). However, 35 days after the primary immunization with TNP-KLH, antigen-specific IgG titers were significantly reduced in KLF2-deficient mice (Fig. 19A, right panel).

The secondary immune response against TNP-KLH (Fig. 19B) mirrored the primary response: over a period of 7 days, the kinetics of antigen-specific IgG and IgM titers did not differ between wildtype and knockout mice; however, total IgG and IgG1 titers, but not IgM titers, were significantly reduced 14 days after secondary boost immunization (Fig. 19B).

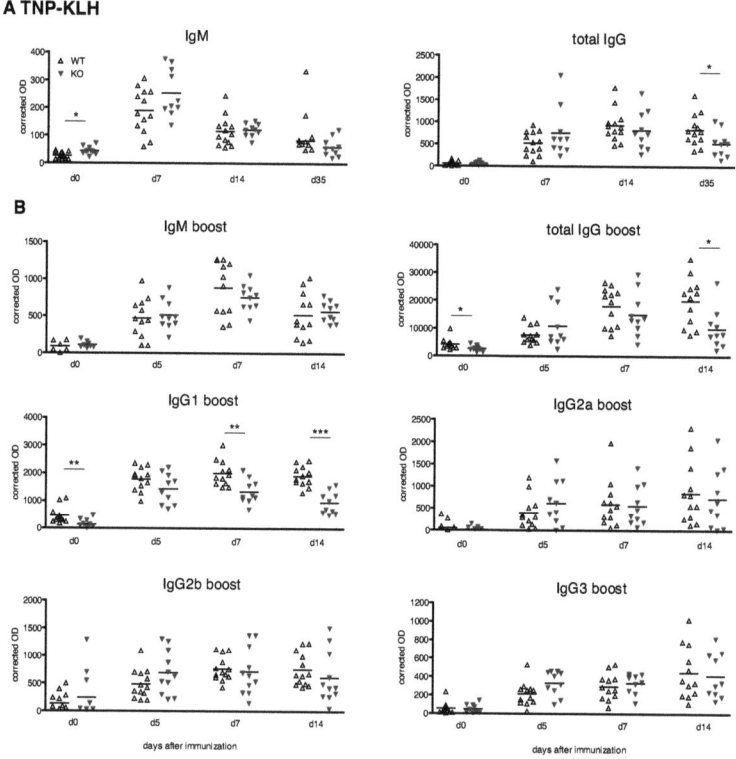

Figure 19. Analyses of Ig-titers after immunization of wildtype and KLF2-deficient mice with the T-dependent antigen TNP-KLH.
Sera of wildtype (blue open triangles) and KLF2-deficient (red filled triangles) mice were collected after primary (**A**: IgM, IgG) and secondary (boost) (**B**: IgM, IgG, IgG1, IgG2a, IgG2b, IgG3) immunization with 100 µg TNP-KLH and analyzed by ELISA for antigen specific antibody production.

This was not due to a drop in the number of antibody-secreting (CD138$^+$/κ$^+$λ$^+$) plasma cells in blood and spleen since their numbers were similar 5 and 14 days after boost immunization in blood and spleen, respectively (Fig. 20A). In contrast, numbers and frequencies of plasma cells in bone marrow were significantly decreased 14 days after secondary immunization with TNP-KLH (Fig. 20A, lower panel). To confirm this finding for antigen-specific plasma cells, we enumerated TNP-specific IgG-secreting cells by Elispot assays in spleen and bone marrow from mice 14 days after boost immunization with TNP-KLH (Fig. 20B). We found similar numbers of IgG producing antigen-specific plasma blasts in the spleen of both wildtype and knockout mice; but in contrast to wildtype mice, TNP-IgG-secreting cells were absent in the bone marrow of KLF2-deficient mice.

Hence, KLF2 contributes to the homing of plasma cells to survival niches in the bone marrow.

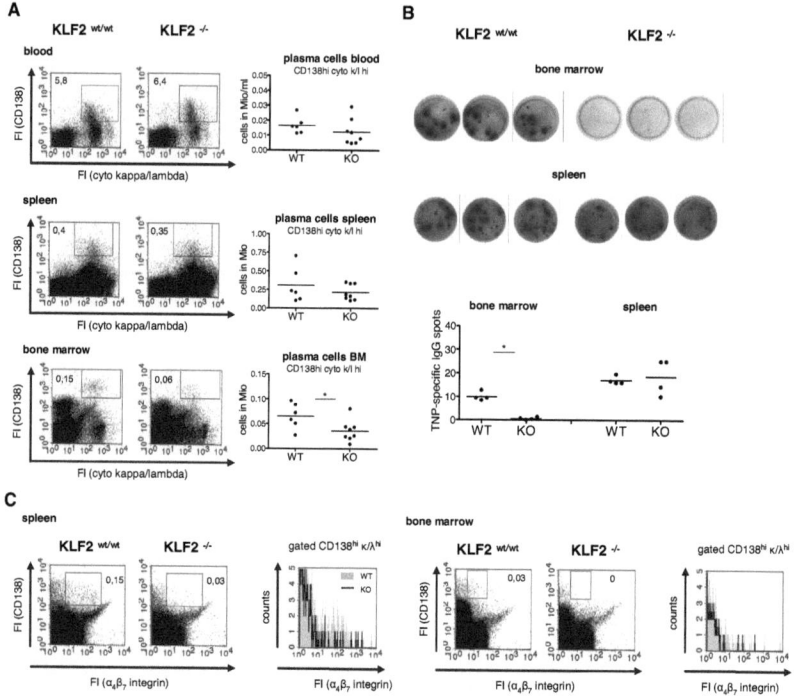

Figure 20. Analyses of plasma cells after immunization of wildtype and KLF2-deficient mice with the T-dependent antigen TNP-KLH.
(A) Flow cytometric analyses of plasma cells (CD138^{hi}/κ/λ^{hi}) in blood (d5), spleen (d14) and bone marrow (d14) of wildtype and KLF2-deficient mice after boost immunization with 100 μg TNP-KLH. FACS dot plots show one representative experiment and scatter plots implement all mice measured and mean values. **(B)** Antigen specific Elispot for IgG producing plasma cells in bone marrow and spleen (d14) after boost immunization with 100 μg TNP-KLH. A triplicate of one representative mouse is shown in the upper panel and the dots in the graph in the lower panel represent the mean value of the triplicates of each mouse (n = 4).

2.1.6 Distribution of peripheral B cell subsets is controlled by KLF2 regulating L-selectin and α$_4$β$_7$ integrin but not S1P1

Since KLF2-deficient animals displayed elevated numbers of B cells in the spleen and reduced numbers of B cells in other peripheral organs, KLF2 might play a role in migration and homing of peripheral B cell subsets. Therefore we investigated the expression profile of S1P1, L-selectin and α$_4$β$_7$ integrin, (gene symbol: *Itgb7*) which contribute to KLF2-dependent migration and homing of T cells (Bai et al., 2007; Carlson et al., 2006; Kaczynski et al., 2003; Sebzda et al., 2008). Egress of thymocytes and B cells from the thymus and the bone marrow, respectively, to the blood is controlled by expression of S1P1, a direct KLF2 target gene (Allende et al., 2010; Carlson et al., 2006; Kuo et al., 1997; Pereira et al., 2010). In addition, S1P1 is required for efficient egress of plasma cells from the spleen into the blood (Kabashima et al., 2006). Since KLF2- deficient B cells egress from the bone marrow and since KLF2-deficient plasma cells can egress from the spleen to the bloodstream, we would expect that KLF2 deficiency, in contrast to T cells, does not significantly affect S1P1 expression. Indeed, this was the case. In TaqMan RT-PCR, microarray and flow cytometric analyses, we detected in splenic Fo B and MZ B cells as well as in blood B cells very similar amounts of S1P1 on protein and transcript levels (Figs. 21A, B and 22A, upper row).

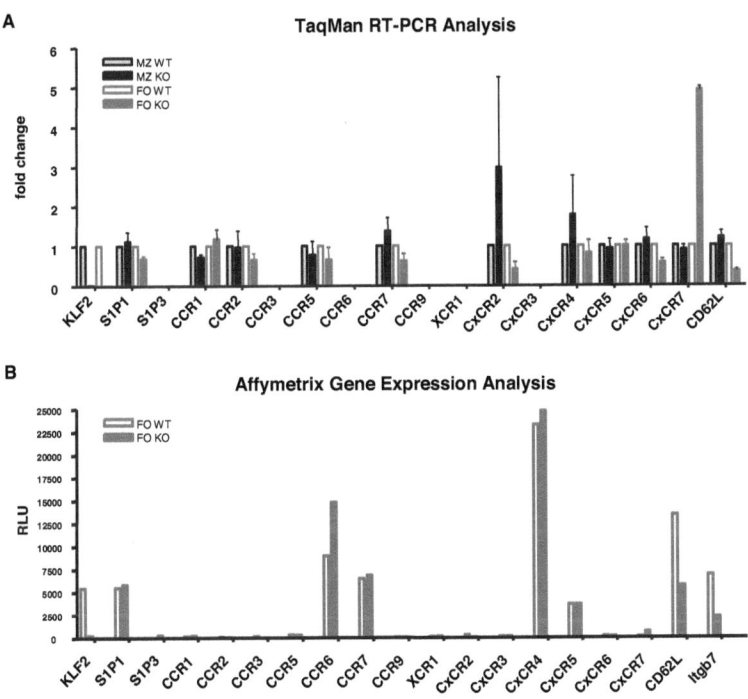

Figure 21. Identification of potential KLF2 target genes.

(A) Quantitative TaqMan PCR of total RNA from primary splenic Fo (red bars) and MZ (black bars) B cells from wildtype (light bars) and KLF2-deficient mice (dark bars) with gene-specific TaqMan probes for selected genes. A TaqMan probe for HPRT was used to calculate fold expression. **(B)** Affymetrix microarray analysis of selected genes in Fo B cells from wildtype and KLF2-deficient mice.

To determine whether S1P1 is still functional on KLF2-deficient MZ B cells, we injected mice with the S1P1-agonist FTY720 (Fig. 22B). FTY720 binding to S1P1 receptor leads to internalization of the receptor. Therefore the cells cannot respond to S1P gradients anymore and as a result, are attracted by the chemokine CXCL13 to the follicular zone (Cinamon et al., 2004). As documented in Figure 22B, FTY720 leads to a displacement of MZ B cells and abolishes the clear separation of Fo B and MZ B cells as indicated by an almost complete overlap of IgM and IgD staining after FTY treatment compared to DMSO treated mice.

There are conflicting reports about the role of KLF2 in controlling expression of chemokine receptors (Sebzda et al., 2008; Weinreich et al., 2009). In our TaqMan RT-PCR and microarray analyses, the expression of most chemokine receptors was not affected in KLF2-deficient Fo und MZ B cells (Fig. 21). Only transcripts for CxCR7 are upregulated in KLF2-deficient Fo B cells compared to their wildtype counterparts on microarrays and TaqMan RT-PCR assays (Figs. 21A, B).

Since $\alpha_4\beta_7$ integrin as well as L–selectin are described KLF2 target genes, we investigated the abundance of these adhesion molecules in KLF2-deficient B cells (Bai et al., 2007; Carlson et al., 2006). We found a reduction of $\alpha_4\beta_7$ integrin in KLF2-deficient mice on recirculating bone marrow B (Fig. 22D), blood B (Fig. 22A), Fo B cells (Figs. 21A, B, 22A, C) and splenic plasma cells (Fig. 22E): here, we measured expression of β_7 integrin (genesymbol: *Itgb7*) mRNA by Affymetrix microarray and TaqMan RT-PCR, and the amount of $\alpha_4\beta_7$ integrin protein was assessed by flow cytometry. In contrast, $\alpha_4\beta_7$ integrin is hardly expressed on wildtype and KLF2-deficient bone marrow plasma cells (Fig. 22E) and MZ B cells (Fig. 22A). This was not surprising since wildtype MZ B cells barely produce KLF2 (Fig. 5D). In conclusion, KLF2-deficiency in B cells results in a strong reduction of $\alpha_4\beta_7$ integrin surface expression.

L-selectin, an adhesion molecule critical for lymphocytes to home to peripheral lymph nodes (Hamann et al., 1994; Wagner et al., 1998) can still be detected at the mRNA and protein level on KLF2-deficient Fo B and blood B cells. But when compared to the corresponding wildtype B cell subsets L-selectin expression was reduced in KLF2-deficient B cells (Fig. 22A, lower row).

In summary, KLF2 does not regulate S1P1 expression in B cells. However, KLF2-deficieny in B cells results in a strong reduction of $\alpha_4\beta_7$ integrin and a weaker downregulation of L-selectin on Fo B cells.

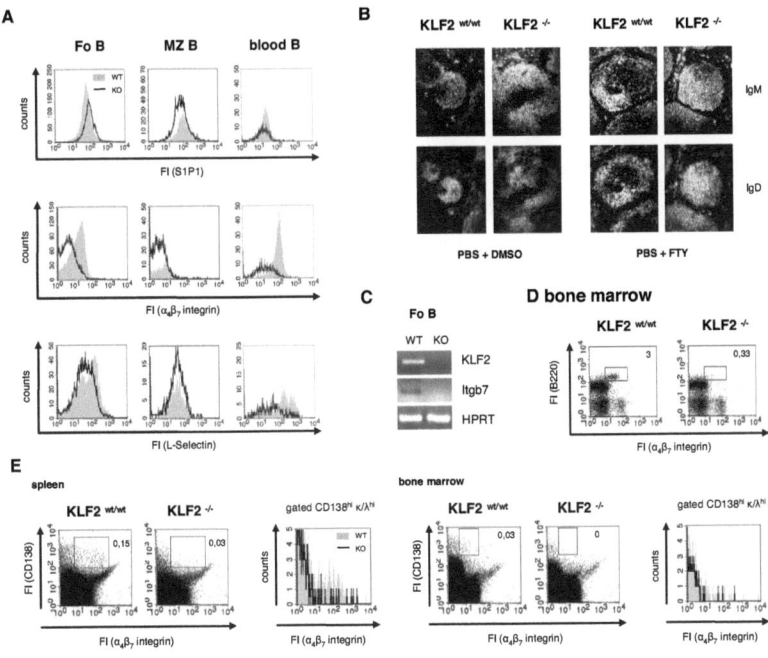

Figure 22. Verification of KLF2 target genes.
(A) Flow cytometric analyses of S1P1, $\alpha_4\beta_7$ integrin and L-selectin in Fo B, MZ B and blood B cells of wildtype (grey histograms) and KLF2-deficient (open histograms) mice. (B) Histological analyses of splenic cryesections from DMSO (left panel) and FTY720 treated (right panel) mice. Acetone fixed tissue cryosections were stained with the indicated fluorochrome conjugated antibodies (stained surface markers are shown to the right). (C) RT-PCR analyses of KLF2 and Itgb7 transcripts in Fo B cells from wildtype and KLF2-deficient mice. HPRT served as a control for the integrity and amount of RNA. (D) Flow cytometric analysis of $\alpha_4\beta_7$ integrin surface expression on recirculating B cells (B220^{hi}) from the bone marrow. (E) Flow cytometric analyses of $\alpha_4\beta_7$ integrin on plasma blasts from the spleen and plasma cells from the bone marrow (d14 after boost immunization). Dot plots show CD138^{hi}/$\alpha_4\beta_7$ integrin$^+$ cells and histograms show the surface expression of $\alpha_4\beta_7$ integrin on gated CD138^{hi}/κ/λ^{hi} plasma blasts in the spleen (to the left) and plasma cells in the bone marrow (to the right). The dot plots and histograms are representative for n=4 wildtype and KLF2-deficient mice.

2.1.7 KLF2 determines Fo B cell identity

Since KLF2 protein is barely detected by western blotting in wildtype MZ B cells (Fig. 5D), we speculated that KLF2-deficient Fo B cells might maintain, at least in part, some molecular and functional hallmarks of MZ B cells. This is indeed the case. As shown by flow cytometry (Figs. 10, second row and 23A) and on splenic cryosections (Fig. 23B), KLF2-deficient Fo B cells fail to downregulate complement receptors 1/2 (CD21/35) and express CD21/CD35 at levels similar to that on MZ B cells. Additionally, CxCR7, a chemokine receptor expressed on MZ B cells (Sierro et al., 2007), is upregulated on KLF2-deficient Fo B cells, as shown by microarray and TaqMan RT-PCR analyses (Figs. 21A, B).

Furthermore, anti-IgM induced Ca^{2+} flux in Fo B cells resulted in a slight but statistically significant increase in the baseline of intracellular Ca^{2+}; and the Ca^{2+} flux pattern in KLF2-deficient Fo B cells resembles more the one found in wildtype and KLF2-deficient MZ B cells (Fig. 23C). Therefore, it was not surprising that MZ-like B cells with higher expression of CD21/35 could also be detected in the inguinal lymph nodes by flow cytometry (Fig. 23D) and on cryosections (Fig. 23E). These findings point to a function of KLF2 in determining the identity of Fo B cells.

The major findings of this part of the thesis were published in the Proceedings of the National Academy of Sciences (Winkelmann et al., 2011).

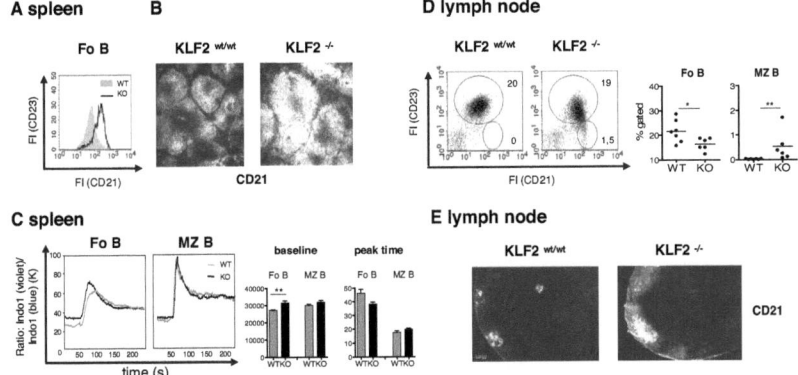

Figure 23. Analyses of complement receptor (CD21/35) expression on spleen and lymph node B cells from wildtype and KLF2-deficient mice.
(A) Flow cytometric analysis of CD21/CD35 expression in Fo B cells of wildtype (grey histograms) and KLF2-deficient (open histograms) mice. (B) Analysis of CD21/35 expression on splenic cryosections stained with a fluorochrome conjugated antibody against CD21. (C) Calcium signaling of Fo and MZ B cells of wildtype (grey line) and KLF2-deficient mice (black line) after IgM treatment. Graphs show one representative experiment and diagrams implement mean values with SEM. (D) Flow cytometric analysis of Fo (B220$^+$/CD23$^+$/CD21^{low}) and MZ (B220$^+$/CD23$^-$/CD21^{hi}) B cells in inguinal lymph nodes. FACS dot plots show one representative experiment and scatter plots implement all mice measured and mean values. (E) Analysis of lymph node architecture. Acetone-fixed tissue cryosections were stained with a fluorochrome conjugated antibody against CD21.

2.2 Biochemical analyses of KLF2

2.2.1 KLF2 protein modifications in splenic B cells and non B cells

Since we detected several specific forms of KLF2 protein in western blot analyses of B cell extracts (Figs. 4B, 5B, C, D, 7C, D) which are differentially downregulated after stimulation (Figs. 5B, C), we suggest that KLF2 is posttranslationally modified and thereby regulated in its activity. To investigate and compare KLF2 modifications in different lymphoid cell populations we used different MACS-sorting strategies to distinguish between splenic B and non B cell populations (T cells, monocytes, stromal cells, etc.). Indeed, also in splenic non B cells different forms of KLF2 could be detected (Fig. 24A), which show even a different pattern compared to B cells. In resting B cells CD19$^+$, CD43$^-$, or IgM$^+$ two prominent forms could be detected whereas in splenic non B cells even more forms with a higher mobility could be found. In fractionizing experiments of NYC or CD43- sorted splenic B cells we could show that the higher molecular weight form is enriched in the nucleus (Fig. 24B). Purity of fractions was controlled by staining with erk1 and Pax5 antibodies for cytosolic and nuclear proteins respectively.

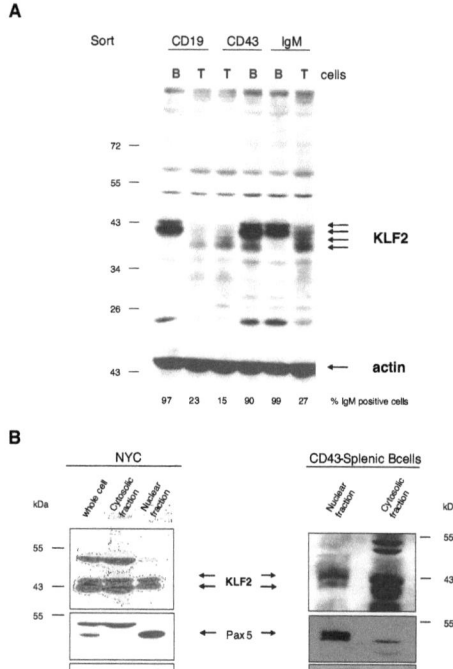

Figure 24. KLF2 expression pattern after different sorting strategies and fractionizing into nuclear and cytosolic proteins.
(A) Primary splenic B cells were MACS sorted with either CD19, CD43 or IgM beads. Positive fractions and negative fractions were analyzed by western blot. CD19+, CD43- and IgM+ fractions contain B cells (labeled as B) and CD19-, CD43+ and IgM- fractions contain T cells, macrophages and stromal cells (labeled as T). Purity of IgM positive cells in the fractions is indicated below. Membranes were stained with an anti-KLF2 antiserum and with an anti-actin serum as loading control. **(B)** NYC and primary splenic B cells were separated in cytosolic and nuclear fractions and analyzed by western blot for KLF2. Membranes were stained with an anti-KLF2 antiserum, a monoclonal anti-Pax5 and -erk1 antibody for control of nuclear and cytosolic fractions respectively.

2.2.2 *In silico* phosphorylation analysis of KLF2 amino acid sequence

Phosphorylation events are important in intracellular signalling cascades. Since KLF2 is expressed after pre-BCR induction and downregulated after BCR stimulation where phosphorylation cascades are activated we investigated *in silico* phosphorylation using the Netphos program. The Netphos program uses a neural network to classify highly complex and non-linear biological sequence patterns. For each acceptor residue tyrosine, threonine and serine a sequence logo was generated emphasizing residues that are frequently found in the context of phosphorylation sites (Blom et al., 1999). *In silico* analysis showed 4 high score potential phosphorylation sites within the proline rich region (2 tyrosine, 1 threonine, 1 serine) and 5 high score serine phosphorylations within the zinc finger domain (Fig. 25) indicating that phosphorylation leads to altered interaction partners (proline rich region) or a different capability of DNA binding (zinc finger region).

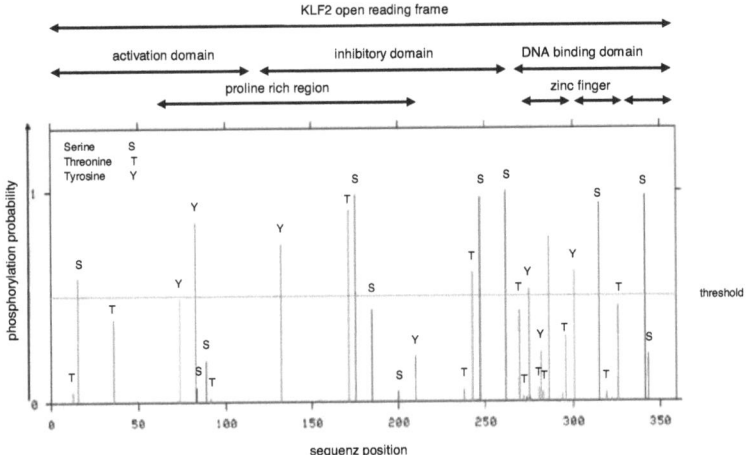

Figure 25. *In silico* analysis of potential KLF2 phosphorylation sites.
Amino acid (AA) sequence of KLF2 was investigated *in silico* using the Netphos program (www.cbs.dtu.dk/services/NetPhos/) for potential phosphorylation sites. Potential phosphorylation sites on serine (blue), threonine (green) and tyrosine (red) are indicated. Threshold is set at 0.5. Known KLF2 domains within the open reading frame are shown upon the Netphosdiagram (www.ensembl.org; proline rich region: 60-215 AA; zinc finger domains: 271-300 AA; 301-330 AS, 331-352 AA).

2.2.3 KLF2 is posttranslationally modified by phosphorylation

With these strong *in silico* indications for KLF2 phosphorylation we treated CD43⁻ splenic B cells with different phosphatase inhibitors *ex vivo*. If KLF2 is phosphorylated, a shift to higher molecular weight forms is expected. We used Pervanadat which blocks tyrosine-/serin- and threonine phosphatases (Vecchi et al., 1998), as well as Calyculin A and Okadaic acid which block serin/threonine kinases (Ishihara et al., 1989; Mumby and Walter, 1993; Vale and Botana, 2008). As shown in Fig. 26A, the highest molecular weight form was strongly enriched in Pervanadat treated samples compared to controls (-, DMSO). In addition, Calyculin A was able to enrich the higher molecular weight forms but less potent than Pervanadat. Treatment with Okadaic acid did not result in a change of the pattern. Furthermore, immunoprecipitation experiments with either precipitation of KLF2 using a polyclonal antiserum and staining with P-tyrosine and -threonine specific antibodies or precipitation of tyrosine and threonine phosphoproteins with subsequent staining using KLF2 specific antiserum revealed, that the higher molecular weight form of KLF2 could be detected in enriched phosphoproteins. Even two forms could be detected after enrichment of KLF2 and staining with anti-phospho-antibodies (Fig. 26B), suggesting that KLF2 is phosphorylated on threonine and tyrosine residues.

PI3 kinase is an important kinase in lymphocyte signalling (Werner et al., 2010), and a study in T cells showed that KLF2 transcripts were upregulated in CD8+ T cells upon PI3K inhibition with either rapamycin or Ly294002 treatment (Sinclair et al., 2008). Therefore we wanted to test whether PI3K is also involved in KLF2 phosphorylation in B cells by inhibiting PI3K using Ly294002. Western blot analyses of Ly294002 treated CD43⁻ splenic B cells revealed that the lower molecular weight form of KLF2 was more prominent compared to DMSO treated control cells (Fig. 26C). Moreover, stimulation for 48h with 10 μg/ml LPS and treatment with Ly294002 diminished LPS induced degradation of KLF2 since both KLF2 forms and especially the lower molecular weight form were more abundant in comparison to DMSO/LPS treated cells.

In summary, we suggest that KLF2 is postranslationally phosphorylated by the PI3K signalling pathway and thereby marked for degradation.

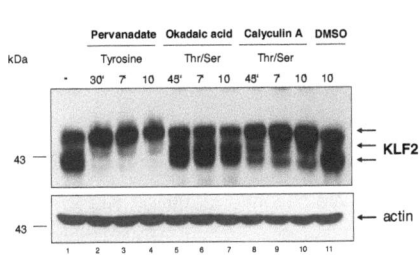

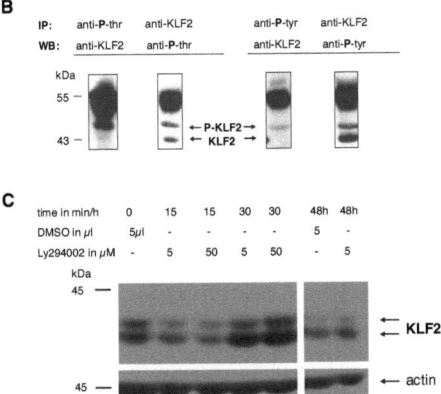

Figure 26. KLF2 is posttranslationally modified by phosphorylation.
(A) Western blot analyses of lysates from 1*10^7 primary CD43-negative B cells with a purity of 88% before and after treatment with either 25 µM Pervanadat, 120 nM Calyculin A or 84 nM Okadaic acid for 30 sec, 7 min and 10 min. Membranes were stained with an anti-KLF2 serum and an anti-actin staining served as loading control. **(B)** Proteins from 2*10^7 NYC cells were precipitated with either phospho-specific antibodies (P-threonine to the left and P-tyrosine to the right) or an anti-KLF2 serum and analyzed by western blot. Membranes were stained with an anti-KLF2 antiserum or P-threonine (left) or P-tyrosine (right) antibodies respectively. **(C)** 1*10^7 CD43$^-$ splenic B cells were treated with 5 or 50 µM of the PI3K inhibitor Ly294002 for 15 min, 30 min, or 48 h (with 10 µg/ml LPS) at 37°C followed by western blot analysis. Membranes were stained with an anti-KLF2 serum and an anti-actin serum as loading control.

3 Discussion

3.1 KLF2-deficiency results in impaired B cells homeostasis and plasma cell homing

KLF2, a member of the krüppel-like transcription factor family, is an established regulator of T cell quiescence and migration. We recently identified KLF2 as a target gene of pre-BCR signalling during early B cell development (Schuh et al., 2008). Microarray studies revealed that KLF2 transcripts are highly abundant in resting B cells, downregulated upon mitogenic activation and reexpressed in plasma and memory B cells (Bhattacharya et al., 2007; Glynne et al., 2000; Kabashima et al., 2006). However, the role of KLF2 in B cells is not fully understood.

In the first part of this thesis we investigated the role of KLF2 deficiency on B cell homeostasis and plasma cell homing. Because of its function as a quiescence factor in T cells, we speculated that KLF2 is a critical regulator of the termination of pre-B cell expansion (Fruman et al., 2002; Glynne et al., 2000; Kuo et al., 1997). Therefore, we expected hyperproliferation of KLF2-deficient pre B cells. However, we only found a slightly increased pre-B cell and immature B cell compartment in the bone marrow and no differences after a single BrdU injection, indicating that loss of KLF2 alone does not result in hyperproliferation. It is very likely that other members of the krüppel-like family, such as KLF4, or other zinc finger transcription factors like Ikaros and Aiolos, which are also expressed in pre-B cells (Klaewsongkram et al., 2007; Ma et al., 2010; Ma et al., 2008; Thompson et al., 2007; Trageser et al., 2009), compensate for the loss of KLF2 (Kharas et al., 2007). Consistent with our finding, bone marrow chimeras with KLF2-deficient T cells displayed only minor effects on proliferation (Bai et al., 2007; Carlson et al., 2006) in contrast to KLF2 overexpressing experiments which lead to induction of cell cycle arrest in T cells (Buckley et al., 2001). In summary, we conclude that KLF2 does not influence central B cell development.

KLF2 deficiency leads to a 2-3fold accumulation of B cells in the spleen, including transitional, Fo, MZ (even up to 5fold) and B1 cells. This accumulation could be due to slightly increased numbers of emigrating immature B cells from the bone marrow, or to defects in the migratory behaviour of Fo and B1 B cells as discussed later.

Positioning of B cells within the spleen is regulated by gradients of the chemokines S1P and CXCL13. S1P1 the receptor for S1P keeps MZ B cells in the marginal zone whereas Fo cells are attracted to follicles via the CxCL13 receptor CxCR5 (Ansel et al., 2002; Cinamon et al., 2004; Cinamon et al., 2008; Vora et al., 2005). KLF2 directly binds to the S1P1 promotor and activates S1P1 expression in T cells (Bai et al., 2007; Carlson et al., 2006). In developing T cells, loss of KLF2 leads to downregulation of S1P1 resulting in impaired egress and accumulation of T cells in the thymus (Carlson et al., 2006; Kuo et al., 1997). In contrast to T cells, we found no change in the expression of S1P1 on B cells as shown by flow cytometry, RT-PCR and Affymetrix microarray analyses, indicating that S1P1 expression in B cells is not solely dependent on KLF2. This notion is supported by our finding that marginal zone B cells express only small amounts of KLF2 but high amounts of S1P1 as reported by the group of J. Cyster (Cinamon et al., 2004). In addition, we could show that S1P1 was still functional since FTY720 treatment displaced MZ B cells from the marginal zone *in vivo* (Cinamon et al., 2004). Furthermore S1P1 is important for egress of immature B cells from the bone marrow (Pereira et al., 2010). Recently it was shown that loss of S1P1 expression results in an accumulation of developing B cells in the bone marrow and, as a consequence, in a reduction of transitional B cells in the spleen (Allende et al., 2010). However, KLF2-deficient animals display normal numbers of immature B cells in the bone marrow and even higher numbers of transitional B cells within the spleen.

We conclude that in contrast to T cells S1P1 expression is not strictly dependent on KLF2 in B cells.

During preparation of my doctoral thesis, Hoek et al. reported the effect of a conditonal KLF2 deletion on B cells (Hoek et al., 2010). While this study also reports an increase in splenic B cell populations and a drastically reduction in peritoneal B1 cells, there are notable differences. Most importantly, the work by Hoek et al. analyzed the effect of B cell-specific KLF2 deficiency on neither secondary immune responses nor plasma cell homing. In addition, the study showed by quantitative SyBr Green RT-PCR and migration assays that S1P1 is drastically downregulated in MZ B cells and upregulated in Fo B cells. In contrast, our study detected no significant changes in S1P1 expression upon KLF2 depletion in MZ and Fo B cells. The drastic downregulation of S1P1 in KLF2-deficient MZ B cells by Hoek et al. is surprising, since MZ B cells barely produce KLF2 protein and still disappear after blockage of

S1P1 function through FTY720 treatment from the marginal zone. Since Hoek et al. did not verify surface expression of S1P1 protein, it still remains to be confirmed whether S1P1 is indeed downregulated at the protein level. However, an independent study conducted by Hart et al. also found no regulation of S1P1 using CD19 cre for KLF2 deletion in B cells (Hart et al., 2011).

Accordingly to observations made in T lymphocytes (Bai et al., 2007; Carlson et al., 2006) we demonstrate that KLF2 is a regulator of L-selectin and $\alpha_4\beta_7$ integrin expression in B cells, since KLF2-deficient B cells display lower surface levels of L-selectin as well as $\alpha_4\beta_7$ integrin. In line with the findings of Weinreich et al. (Weinreich et al., 2009) we could not find significant alterations of chemokine receptor levels in KLF2-deficient B cells.

The downregulation of L-selectin and $\alpha_4\beta_7$ integrin could also explain defective B cell homeostasis in KLF2-deficient mice, since the phenotype is mirrored in mice with defective L-selectin and/or $\alpha_4\beta_7$ integrin function. For example, treatment with blocking antibodies against L-selectin and/or $\alpha_4\beta_7$ integrin (Hamann et al., 1994) or germline deletion of β_7 integrin (Wagner et al., 1998) results in an accumulation of B cells in the spleen and a reduced capacity of B cells to enter peyers patches and mesenteric lymph nodes. Moreover, loss of B cells in peyer patches was accompanied by low IgA serum levels. As mucosal immunity is critically dependent on L-selectin and $\alpha_4\beta_7$ integrin expression (Cyster, 2003; Mora and von Andrian, 2008) we conclude that KLF2 plays a role in proper assembly of gut-associated lymphoid tissues by enhancing B cell migration to mucosal sites.

Peritoneal B1a and B1b cells were almost absent in KLF2-deficient animals. Surprisingly this lack of B1 cells is not accompanied by decreased serum IgM levels (Berland and Wortis, 2002). A possible explanation could be the 4-5fold increased marginal zone B cell or the slightly increased B1a cell compartment in the spleen which might compensate the loss of the peritoneal B1 cells as a producer of "natural" IgM (Holodick et al., 2010; Martin and Kearney, 2002). Alternatively KLF2-deficient B1 cells might be located outside the peritoneal cavity, e.g. in the omentum or the parathymic lymph nodes (Kunisawa et al., 2007). Since $\alpha_4\beta_7$ integrin was described to be important for B2 cell trafficking through milky spots into the peritoneal cavity (Berberich et al., 2008), it is possible that B1 cells also need $\alpha_4\beta_7$ integrin to reach the peritoneal cavity. Therefore the loss of $\alpha_4\beta_7$ integrin in KLF2-deficient mice could explain the observation that B1 cells can not reach the peritoneal cavity anymore and accumulate in the spleen.

Taken together, we identified KLF2 as an upstream regulator of L-selectin and $\alpha_4\beta_7$ integrin in B cells leading to dysregulated B cell homeostasis in KLF2-deficient mice.

Since KLF2 is barely detected in wildtype MZ B cells and loss of KLF2 leads to an expansion of the MZ B cell compartment we speculated that KLF2 might maintain some molecular and functional hallmarks of MZ B cells. Indeed we could show that KLF2-deficient Fo B cells have higher amounts of CD21/35 protein on their surface compared to wildtype Fo B cells. These receptors are important for the recognition of blood borne antigens (Martin and Kearney, 2002) and normally highly expressed on MZ B cells. In addition KLF2-deficient Fo B cells express more CxCR7 similar to MZ B cells (Sierro et al., 2007). By analyzing calcium signals in response to anti-IgM treatment we could find that KLF2-deficient Fo B cells mount a calcium signal that resembles that of MZ B cells. Moreover, we detected abnormal MZ like cells, as defined by $CD23^{lo}/CD21/35^{hi}$ expression, in inguinal lymph nodes.

In summary, our findings indicate that KLF2 is a novel determinator of Fo B cell identity; KLF2 limits the expression of CD21/35 and CxCR7 and modulates the calcium response upon BCR stimulation by so far unknown mechanisms.

It is known that plasma cells require CxCR4 and CxCR3 to reach the bone marrow (Hauser et al., 2003; Nie et al., 2004). But how they enter the bone marrow and get in contact with cells providing survival factors such as IL-6 (Minges Wols et al., 2002), BAFF/Blys (O'Connor et al., 2004), IL-5, CxCL12 (Nie et al., 2004), TNF and CD44 ligands (Cassese et al., 2003) is not clear (Tarlinton et al., 2008; Tokoyoda et al., 2010). Here we show that loss of KLF2 leads to a significant reduction of plasma cells in the bone marrow, indicating that KLF2 regulates factors for adhesion or chemokine receptors important for PC homing. Since we found the same numbers of plasma cells in the spleen, the generation of PCs seems not to be affected by the loss of KLF2. Moreover, the egress of PCs from the spleen to the blood which is dependent on S1P1 (Kabashima et al., 2006) seems not to be impaired since we found no accumulation of PCs in the spleen of KLF2-deficient animals and no significant difference in the numbers of PCs in the blood. We therefore speculate that $\alpha_4\beta_7$ integrin regulated by KLF2 is a novel homing factor for PCs to the bone marrow. In line with the findings that $\alpha_4\beta_7$ integrin together with CCR10 (all mucosal compartments) and CCR9 (small intestine) are important for IgA ASCs to home to mucosal tissues (Mora and von Andrian, 2008), $\alpha_4\beta_7$ integrin could be the integrin helping CxCR4 to direct ASCs to the bone marrow. This is further supported by the notion that myeloma cells can bind VCAM-1 which is in combination with MAdCAM-1 a ligand for $\alpha_4\beta_7$ integrin and highly expressed in the bone marrow (Cyster, 2003). Moreover, recirculating B cells normally express high levels of $\alpha_4\beta_7$ integrin on their surface. Since we found a reduction in the number of recirculating B cells in the bone marrow of KLF2-deficient animals, we conclude that not only plasma cells but also recirculating B cells depend on KLF2 and subsequent expression of $\alpha_4\beta_7$ integrin to enter the bone marrow.

We propose that $\alpha_4\beta_7$ integrin regulated by KLF2 is one key molecule for trafficking of B cells, since circulating B cells need $\alpha_4\beta_7$ integrin to enter the bone marrow, peyers patches, mesenteric lymph nodes and the peritoneal cavity. In contrast they are passively released in sinuses of the marginal zone and red pulp in the spleen (Stein and Nombela-Arrieta, 2005) which leads to an accumulation of B2 and B 1 cells in the spleen and a reduction in nearly all other compartments where B cells circulate. One exception are peripheral lymph nodes where B cell frequencies are normal in KLF2-deficient mice. Here we propose that only L-selectin is important (Okada et al., 2002) which is only weakly downregulated after KLF2 deletion.

Hence, we suggest KLF2 being a master regulator of B cell homeostasis, follicular B cell identity and plasma cell homing to survival niches.

3.2 Biochemical analyses of KLF2

In the second part of this thesis we investigated posttranslational modifications of KLF2 since several forms of KLF2 were detected in western blot analyses. We could show that KLF2 is posttranslationally phosphorylated under involvement of the PI3K signalling pathway. Three findings led to this conclusion: Firstly, KLF2 higher molecular weight forms were enriched after phosphatase inhibitor treatment. Secondly, precipitated KLF2 could be detected with phospho-tyrosine/threonine specific antibodies and *vice versa* and thirdly, stimulation with the PI3 kinase inhibitor Ly294002 led to an enrichment of the lower molecular weight form. Since the higher molecular weight form vanishes earlier after antigen stimulation, we suggest that phosphorylation of KLF2 leads to its ubiquitiniation and thereby degradation as described earlier (Fabre et al., 2008) with direct or indirect involvement of PI3K. This suggestion is supported by the doctoral thesis of P. Dießenbacher, showing that phosphorylation of human KLF2 indeed leads to instability of the protein (Dießenbacher, 2005).

The model in Fig. 27 integrates the findings into a biological context. After activation of the B cell receptor PI3K signalling is induced and phosphorylates proteinkinase B (Akt, PKB) which then phosphorylates Foxo1 leading to cytosol retention in a complex with 14-3-3 proteins and Foxo1 inactivation (Brunet et al., 1999). Foxo1 binds to and activates the KLF2 promotor under resting conditions (Fabre et al., 2008) and phosphatases inhibit KLF2 degradation resulting in a steady state expression of KLF2. After nuclear export of P-Foxo1, *de novo* synthesis of KLF2 is blocked. Furthermore, BCR activation leads to induction of other kinases which lead to predominant KLF2 phosphorylation and nuclear export, resulting in proteasomal degradation. This model fits also well the situation found in pre-B cells. After pre-BCR induction PI3K is activated and leads to nuclear export and inactivation of Foxo1 via Akt. RAG expression is switched off and pre-B cells start clonal expansion (Werner et al., 2010). However, SLP-65 inhibits PI3K activity via yet unknown mechanisms (Herzog et al., 2008). Foxo1 is reexpressed and can induce expression of target genes, such as RAG1 and 2 as well as KLF2, which is upregulated as a late target gene of the pre-BCR and highly abundant in small resting pre-B cells (Schuh et al., 2008).

Since an interaction with the ubiquitin ligase WWP1 is reported (Conkright et al., 2001; Zhang et al., 2004) it is possible that WWP1 or another E3 ligase binds the proline rich region of KLF2 after phosphorylation and marks ubiquitinated, phosphorylated KLF2 for degradation via the proteasome. Furthermore, there are several possible phosphorylation sites mapped in the zinc finger domains which could lead to structural alterations and an impairment of DNA binding.

In summary we suggest that KLF2 posttranslational phosphorylation via the PI3K signalling pathway leads to degradation of the protein via the proteasome.

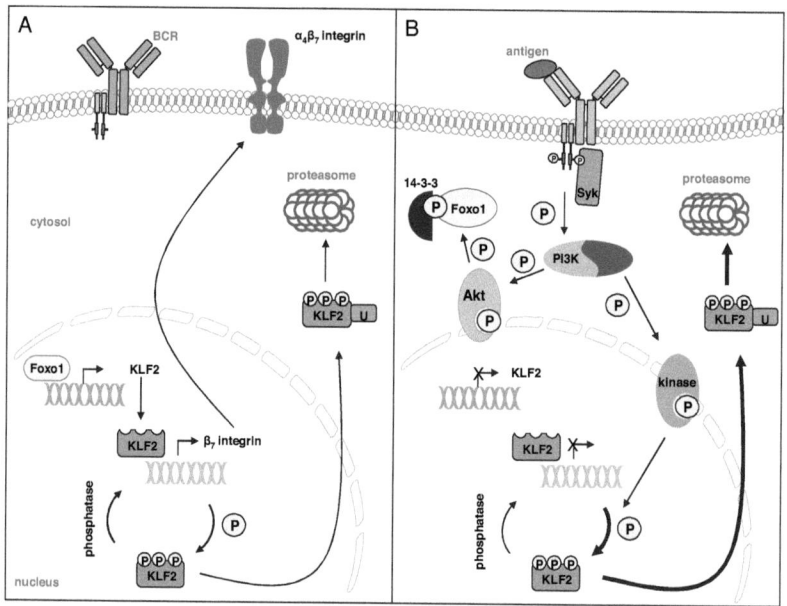

Figure 27. KLF2 phosphorylation via PI3K signalling leads to its degradation.
(A) In resting B cells Foxo1 promotes KLF2 expression. KLF2 steady state abundance is controlled by equilibrium between degradation via phosphorylation and dephosphorylation, leading to transcription and cell surface presentation of $\alpha_4\beta_7$ integrin. **(B)** After antigen encounter the BCR activates the PI3K signalling pathway. Akt kinase gets activated and phosphorylates Foxo1. P-Foxo1 is exported from the nucleus and kept inactivated in the cytosol in a complex with 14-3-3, which blocks KLF2 *de novo* synthesis. PI3K furthermore phosphorylates a yet unknown "kinase" which shifts the equilibrium between phosphorylation and dephosphorylation of KLF2 towards phosphorylation. This leads to nuclear export and proteasomal degradation of KLF2. Therefore KLF2 can no longer activate expression of $\alpha_4\beta_7$ integrin.

In this study we could show that KLF2 plays indeed an important role in B cell homeostasis and PC homing and is strictly regulated during B cell development on the transcriptional and posttranslational level. Further investigations should focus on the molecular mechanisms underlying defective B cell homeostasis and plasma cell homing. For example the investigation whether peritoneal B1 cells do still develop or have defects in their survival or migration capacity in KLF2-deficient mice. The Blimp-GFP mouse model could be used to follow PCs on their way to survival niches and could also be helpful to examine whether KLF2 has an impact on B cell memory. Furthermore, the question why KLF2 regulates S1P1 expression in T cells but not in B cells remains interesting. Here, the occurrence of different KLF2 forms should be analyzed in more detail as well as differences in the KLF2 interactom of T and B cells.

Since KLF2 has so many important functions in T as well as in B cells, like cell cycle control, migration and homeostasis, it remains to be addressed whether KLF2 plays a role in lymphomagenesis and the onset of auto immune diseases.

4 Material and Methods

4.1 Material

4.1.1 Chemicals

[γ-^{32}P]ATP, 10mCi/ml	Hartmann
Agarose	peqlab
Ammonium persulfate (APS)	Sigma-Aldrich
Ampicillin	Sigma-Aldrich
BSA 10X	New England Biolabs (NEB)
p-cumaric acid	Sigma-Aldrich
Desoxyribonucleotides	Gennaxon
EDTA	Sigma-Aldrich
Ethidium bromide (10mg/ml)	Sigma-Aldrich
Ficoll (1077g/ml)	Sigma-Aldrich
GeneRuler™ 100bp-plus DNA ladder	MBI-Fermentas
GeneRuler™ 1kb-plus DNA ladder	MBI-Fermentas
Glycine	Sigma-Aldrich
D-Luminol	Sigma-Aldrich
Magnesium chloride	Sigma-Aldrich
peqGOLD TriFast	peqlab
peqlab GOLD prestained protein standard	peqlab
Hexadimethrinbromid (Polybren)	Fluka
TEMED	Sigma-Aldrich
Tris	Sigma-Aldrich
Trypan blue	Sigma-Aldrich

All other chemicals were obtained from Merck, Carl Roth or GibcoBRL/Invitrogen.

4.1.2 Antibodies

Antibodies are listed in alphabetical order. Antibody dilutions for different experimental approaches are indicated. WB, western blot; IF, immunofluorescence; FACS, flow cytometry.

Table 2. Antibodies in alphabetical order
Abbreviations: AF647: alexa Fluor 647; APC: allophycocyanin; FACS: fluorescence activated cell sorting; FITC: Fluoresceinisothiocyanat; HRP: Horse-raddish-peroxidase; PE: Phycoerythrin; WB: western blot, IF: immunofluorescence, IC: intracellular.

antigen	antibody	dilution	clone	source
actin	rabbit-Ig-anti-mouse actin	WB 1:1000		Sigma-Aldrich
B220	PerCP-rat IgG-anti-mouse B220	FACS 1:200	RA3-6B2	BD
biotin	Cy5-streptavidin	FACS 1:800		Jackson
biotin	PerCP-streptavidin	FACS 1:2000		BD
biotin	Cy3-streptavidin	IF 1:1000		GE Healthcare

antigen	antibody	dilution	clone	source
BP-1	FITC-rat IgG-anti-mouse BP-1	FACS 1:200	6C3	Biolegend
c-kit	APC-rat IgG-anti-mouse c-kit	FACS 1:100	2B8	BD
c-kit	PE-rat IgG-anti-mouse c-kit	FACS 1:100	ACK45	BD
CCR7	biotin-rat IgG-anti-mouse CCR7	FACS 1:40	4B12	eBioscience
CD3	PE-hamster IgG-anti-mouse CD3e	IF 1:100	145-2C11	BD
CD4	FITC-rat IgG-anti-mouse CD4	FACS 1:400	GK1.5	BD
CD4	PE-rat IgG-anti-mouse CD4	FACS 1:800	GK1.5	BD
CD5	PE-rat IgG-anti-mouse CD5	FACS 1:100	53-7.3	BD
CD8	FITC-rat IgG-anti-mouse CD8	FACS 1:400		BD
CD8	APC-rat IgG-anti-mouse CD8	FACS 1:800	53-6.7	eBioscience
CD11b	FITC-rat IgG-anti-mouse CD11b	FACS 1:100	M1/70	eBioscience
CD16/32	FITC-rat IgG-anti-mouse CD16/32	FACS 1:200	FCR4G8	Serotec
CD19	AF647-rat IgG-anti-mouse CD19	FACS 1:1500	eBio1D3	eBioscience
CD19	FITC-rat IgG-anti-mouse CD19	FACS 1:400	1D3	BD
CD19	PE-rat IgG-anti-mouse CD19	FACS 1:400	eBio1D3	eBioscience
CD19	PE-rat IgG-anti-mouse CD19	FACS 1:200	1D3	BD
CD21	FITC-rat IgG-anti-mouse CD21	FACS 1:800	eBio8D9	eBioscience
CD21	FITC-rat IgG-anti-mouse CD21	FACS 1:200	8D9	BD
CD21	biotin-rat IgG-anti-mouse CD21	FACS 1:100	eBio8D9	eBioscience
CD23	PE-rat IgG-anti-mouse CD23	FACS 1:200	B3B4	BD
CD23	PE-rat IgG-anti-mouse CD23	FACS 1:800	B3B4	eBioscience
CD24	Cy5-rat IgG-anti-mouse CD24	FACS 1:1500	M1/69	Biolegend
CD25	PE-rat IgG-anti-mouse CD25	FACS 1:100	PC61.5	BD
CD25	PE-rat IgG-anti-mouse CD25	FACS 1:100	PC61.5	eBioscience
CD43	FITC-rat IgG-anti-mouse CD43	FACS 1:200	S7	BD
CD93	biotin-rat IgG-anti-mouse CD93	FACS 1:100	AA4.1	eBioscience
CD138	APC-rat IgG-anti-mouse CD138	FACS 1:500	281-2	BD
CD138	biotin-rat IgG-anti-mouse CD138	FACS 1:100	281-2	BD
Erk1	Mouse monoclonal anti-erk1	WB 1:500	G262-118	BD
IgA	goat anti-mouse IgA (α-chain specific)	ELISA 1:1000		Southern Biotec
IgA	HRP-goat anti-mouse IgA (α-chain specific)	ELISA 1:2000		Southern Biotec

antigen	antibody	dilution	clone	source
IgD	biotin-rat IgG-anti-mouse IgD	FACS 1:1000 IF 1:200	1126	Southern Biotec
IgD	FITC-rat IgG-anti-mouse IgD	FACS 1:200 IF 1:200		Southern Biotec
IgG	HRP-donkey anti goat IgG	WB 1:100000		Santa Cruz
IgG	goat anti-mouse IgG (γ-chain specific)	ELISA 1:1000		Southern Biotec
IgG	HRP-goat anti-rabbit IgG (H+L)	WB 1:10000		Biorad
IgG	HRP anti-mouse IgG (Fc-fragment specific)	WB 1:10000		Jackson
IgG	HRP-goat anti-rabbit IgG	WB 1:10000		Southern Biotec
IgG	HRP-goat anti-mouse IgG (γ-chain specific)	ELISA 1:2000		Southern Biotec
IgG1	HRP-goat anti-mouse IgG1 (γ1-chain specific)	ELISA 1:2000		Southern Biotec
IgG2a	HRP-goat anti-mouse IgG2a (γ2a-chain specific)	ELISA 1:2000		Southern Biotec
IgG2b	HRP-goat anti-mouse IgG2b (γ2b-chain specific)	ELISA 1:2000		Southern Biotec
IgG3	HRP-goat anti-mouse IgG3 (γ3-chain specific)	ELISA 1:2000		Southern Biotec
IgM	goat-anti-mouse IgM (μ-chain specific)	ELISA 1:1000		Southern Biotec
IgM	HRP-goat anti-mouse IgM (μ-chain specific)	ELISA 1:2000		Southern Biotec
IgM	FITC-goat anti-mouse IgM (μ-chain specific)	FACS 1:400		Southern Biotec
IgM	Cy5-goat anti-mouse IgM (μ-chain specific)	FACS 1:400, IF 1:400		Southern Biotec
Itgb7	biotin-rat IgG-anti-mouse LPAM-1	FACS 1:100	DATK32	eBioscience
kappa	PE-rat IgG-anti-mouse kappa	FACS 1:500 IC	187	Southern Biotec
lambda	PE-rat IgG-anti-mouse lambda	FACS 1:500 IC		Southern Biotec
mouse IgG	HRP-goat Ig-anti-mouse IgG (H+L)	WB 1:150000		Jackson
Madcam-1	biotin-rat IgG-anti-mouse Madcam1	IF 1:100	MECA 367	eBioscience

antigen	antibody	dilution	clone	source
MOMA	biotin-rat IgG-anti-mouse MOMA 1	IF 1:100	MOMA-1	BMA Biomedicals
NK1.1	PE-mouse IgG-anti-mouse NK1.1	FACS 1:200	PK136	BD
Pax5	goat anti-mousel Pax 5	WB 1:200		Santa Cruz
PNA	FITC IgG-anti-mouse PNA	IF 1:400	FL1071	Vector lab
P-Threonin	Polyclonal rabbit anti-P-Thr antibody	WB 1:1000		Cell Signalling
P-Tyrosin	HRP-PY99	WB 1:4000		Santa Cruz
S1P1	polyclonal rabbit anti-human/mouse-S1P1	FACS 1:500		Cayman Chemicals

4.1.3 PCR primer

Shown are oligonucleotides used in PCR, RT-PCR and sequencing reactions, as well as oligonucleotide probes, syntethized by Invitrogen. 5' to 3' sequences are listed according to ER-numbers (internal laboratory indexing).

Table 3. Oligonucleotides

number	name	sequence (5'-3')
2153	Mb1-cre_for	CCCTGTGGATGCCACCTC
2154	Mb1-cre_back	GTCCTGGCATCTGTCAGAG
2171	LKLF_KO_for	ACTTTCGCCAGCCCGTGCGAGCG
2172	LKLF_KO_back	TGAATTCTCGGCGCCCAGACCGTCC
2173	CreI_PM_for CD19	GTTCGCAAGAACCTGATGGACA
2174	CreI_PM_back CD19	CTAGAGCCTGTTTTGCACGTTC
2465	CxCR2 for	GCTCACAAACAGCGTCGTAG
2466	CxCR2 back	CCACAAGGCTCAGCAGAGTC
2467	CxCR3 for	AGCCAAGCCATGTACCTTGAG
2468	CxCR3 back	TCAGGGCAGTGCGCTGACTC
2469	CxCR5 for	ATGGATTTTCAAGGGTCAGT
2470	CxCR5 back	CCAAGTACCTATCAATTGTC
2471	CxCR7 for	AAGGAGCCTGCAGCGCTCAC
2472	CxCR7 back	TCCCAAAGAGGTTGATGGAG
2473	XCR1 for	GTACAGACTTGAAACCCTG
2474	XCR1 back	GACACAGGTTGAGGATGAAG
2475	CCR1 for	GTTGGGACCTTGAACCTTGA
2476	CCR1 back	CCCAAAGGCTCTTACAGCAG

number	name	sequence (5'-3')
2477	CCR3 for	AAACTTGCAAAACCTGAGAAGC
2478	CCR3 back	CATGACCCCAGCTCTTTGAT
2479	CCR5 for	GCCAGAGGAGGTGAGACATC
2480	CCR5 back	CCCACAAAACCAAAGATGAA
2481	CCR6 for	GGCTCTCCCATCCACATAGA
2482	CCR6 back	AAGGCAAAGGTCATCACCAC
2483	CCR7 for	AAAGCACAGCCTTCCTGTGT
2484	CCR7 back	CCACGAAGCAGATGACAGAA
2817	Hcre dir_SM	ACCTCTGATGAAGTCAGGAAGAAC
2818	Hcre rev_SM	GGAGATGTCCTTCACTCTGATTCT
2819	Mb1 dir_SM	CTGCGGGTAGAAGGGGGTC
2820	Mb1 rev_SM	CCTTGCGAGGTCAGGGAGCC

4.1.4 Plasmids

Table 4. Established plasmids

number	name	resistance	reference
E1300	pBMN_KLF2-IRES-GFP	amp, puro	Wolfgang Schuh
E1329	pBMN-IRES-GFP	amp, puro	Jörg Kirberg

4.1.5 Cell lines

Phoenix-eco	derivate of HEK293 cell line adenovirally transfected with retroviral gag, env and pol (Pear et al., 1993)
NIH3T3	contact inhibited NIH swiss mouse embryo fibroblast cell line obtained from ATTC (Jainchill et al., 1969)
38B9	Abelson transformed pro-B cell line (Alt et al., 1984)
TK-	Ableson transformed mouse pre-B cell line (Keyna et al., 1995)
TKµ	µ-transfected, Abelson transformed pre-B cell line (Keyna et al., 1995)
WEHI 279	immature B cell lymphoma cell line (Warner, 1974)
WEHI 231 (FS/CP)	immature B cell lymphoma cell line. (CP or FS means cultivation in different laboratories) (Gutman et al., 1981).
CH27	B cell lymphoma generated after transfer of splenic cells from mice that were hyperimmunized with SRBC (Haughton et al., 1986)
NYC31	B lymphoma isolated from a tumor of NZBxNZW spleen by Nancy Y Cheng
Raji	EBV positive human Burkitt lymphoma cell line from T. Ellis
Ramos	EBV negative B lymphoblast line from Greg Spears
Jurkat	human T cell lymphoma (CD4+, CD30+, TCR+, CD3+, CD8-) (ATCC TIB-152)

4.1.6 Mouse strains

All animals were kept in the Franz Penzoldt Center animal facility, Erlangen. Transgenic animals were genotyped by PCR using total DNA from tail biopsies. Male and female mice were used for experiments.

Table 5. Mouse strain overview

genotype	reference
C57 Bl/6 NRj	Janvier
Balb/c JRj	Janvier
mb1-Cre	(Hobeika et al., 2006)
CD19 cre	(Schmidt-Supprian et al., 2007)
KLF2 flox/flox	(Lee et al., 2006)

4.1.7 Bacteria

E. coli GeneHogs: F- mcrA Δ(mrr-hsdRMS-mcrBC) φ80lacZΔM15 ΔlacX74 recA1 araD139 Δ(araleu) 7697 galU galK rpsL (StrR) endA1 nupG fhuA:IS2 were used for all cloning experiments. For amplification of recombinant plasmids the DH5α strain was more suitable: DH5α: F⁻, ø80dlacZΔM15, Δ(lacZYA-argF)U169, deoR, recA1, endA1, hsdR17(rk⁻, mk⁺), phoA, supE44, λ⁻, thi-1, gyrA96, relA1

Bacterial suspensions were cultivated in LB-medium (10 g Bacto Trypton, 5 g yeastextract, 5 g sodium chloride, ad 1l dH$_2$O) overnight at 37°C in a shaker or plated as single colonies on LB-plates in an incubator (10 g Bacto Trypton, 5 g yeastextract, 5 g sodium chloride, 12 g Agar-Agar, ad 1l dH$_2$O). For selection of ampicillin resistant clones ampicillin was added in a final concentration of 100 µg/ml.

4.1.8 Technical Equipment

Table 6. Technical equipment in alphabetical order

equipment	source
Bunsen burner: Fireboy	Techomara AG
Centrifuge Megafuge 1.0 R	Heraeus
Centrifuge Biofuge fresco	Heraeus
Centrifuge Biofuge pico	Heraeus
Cytometer LSR II	BD Biosciences Biosciences
Cytometer FACS Calibur	BD Biosciences Biosciences
Deionization equipment (Purelab Plus)	USF
Electrophoresis chamber for PAA Gels, Hoefer	SERVA Electrophoresis GmbH
Electro blotter semi-dry	peqlab
Electroporation equipment	Bio-Rad Laboratories

Gel documentation transilluminator	Herolab
Gene amp PCR System 9700	Applied Biosystems
Incubator: CO_2-Autozero	Heraeusu
Luminometer Sirius	Berthold Detection Systems
Magnetic stirrer: IKAMAG REO	IKA Labortechnik
Microscop Axiostar	Zeiss
Nanodrop ND-1000 spectrometer	peqlab
Orbital shaker Red Rotor	Hoefer
PerfectBlue™ –System	peqlab
Pipetboy	Gilson
Power supply unit PS 250 Hybaid	MWG Biotech
7300 Real Time PCR System	Applied Biosystems
scale Lab Style 3002	Mettler TOLEDO
scale Lab Style 303	Mettler TOLEDO
Suction pump type GA-H	Dürr Technik
Thermoshaker comfort	Eppendorf
UV lamp: MacroVue UV-25	Hoefer
Vortexer: Genie 2	Bender & Hobein AG
Water bath, mit Deckel	Memmert
Water bath, offen (Julabo U3)	Julabo Labortechnik

4.1.9 Plastic material

Table 7. Plastic material in alphabetical order

name	source
Filter, bottle top 0,45 µm or 0,22 µm	Becton Dickinson
sterile filter Millex-GV4, 0,22 µm	Millipore
Injection cannula (0,8 x 40 mm, sterile)	Becton Dickinson
Kapillarspitzen, 200µl	BIOzym Diagnostik
Optical 96 well reaction plate (MicroAmp)	ThermoFisher Scientific
Optical adhesive cover sheeting	ThermoFisher Scientific
Pasteur pipettes, 230 mm	Carl Roth
PCR soft strips 0,2 mL	BIOzym Diagnostik
syringe, sterile: 2 mL, 5 mL, 10 mL, 20 mL	Becton Dickinson
Weighing pan	Carl Roth

All other plastic material was obtained from Greiner and Sarstedt.

4.1.10 Consumable material

Table 8. Consumable material in alphabetiacl order

name	source
Autoradiography film (Hyperfilm MP)	GE Healthcare
Autoradiography film cassette (*FBXC 1417*)	ThermoFisher Scientific
Cover slips, 18x18, 24x50 mm	Menzel-Gläser
Glas bottles (Schott, Duran)	Roth
Glas jars (Schott, Duran)	Roth
Magnetic stirring bar	neoLab
Nitrocellulose membrane	Scheicher&Schuell
Parafilm M	American National Can
Kimwipes	Kimberly-Clark
Weighing paper MN226	Macherey-Nagel
Neubauer counting chamber: 0,1 mm x 0,0025 mm^2	Laborcenter

4.2 Methods

4.2.1 Cell culture

4.2.1.1 Cultivation and harvest of vertebrate cell lines

Suspension cells were cultured in RPMI 1640 supplemented with 5% (R5) or 10% (R10) heat inactivated fetal calf serum (FCS), 50 U/ml penicillin, 50 µg/ml streptomycin, 2 mM L-glutamin, 1 mM pyruvate and 50 µM β-mercaptoethanol at 37°C and 5% CO_2. Adherent Phoenix-Eco cells (Pear et al., 1993) for production of retroviral supernatants, murine NIH3T3 fibroblasts, as well as the human HEK293 (*human embryonic kidney*) cell line, were cultured in Dulbecco´s Modified Eagle medium (DMEM) supplemented with 10% (D10) heat inactivated fetal calf serum, 50 U/ml penicillin, 50 µg/ml streptomycin, 2 mM L-glutamin und 1 mM pyruvate at 37°C and 7.5% CO_2.

For all experiments cells were kept under sterile conditions. Cells kept in suspension were divided in dependency of cell densitiy. Adherend cells, Phoenix-Eco, NIH3T3 and HEK293 cells, were washed with sterile PBS and detached with sterile Trypsin-EDTA (0.05%) for a maximum of 5 min at 37°C. The reaction was stopped with 4 times the appropriate cell culture medium and cells were centrifuged at 1500 rpm for 5 min and 4°C (300 g, Heraeus Megafuge). For further cultivation the cell pellet was suspended in cell culture medium and divided into cell culture flasks. For experiments, cells were washed in sterile PBS and kept on ice.

4.2.1.2 Cryoconservation

For longtime storage 5×10^6 cells were centrifuged 5 min at 1500 rpm and 4°C and suspended in 1.5 ml freezing medium (90% FCS, 10% DMSO). Afterwards cells were slowly frozen in an isopropanol storage cask at -70°C. Final storage was performed in liquid nitrogen (-180°C).

4.2.1.3 Thawing of cryoconserved cells

After thawing in a water bath at 37°C, freezing medium was diluted with 20 ml of the appropriate cell culture medium. Cells were centrifuged for 5 min at 1500 rpm and 4°C. Supernatant was removed and cells suspended in cell culture medium for seeding.

4.2.1.4 Cell counting

Living cell numbers were calculated by trypan blue staining using the *Neubauer-Zählkammer*. Cells were diluted with the same volume of trypanblue in PBS [4% (m/v)] (dilution factor of 2). Trypan blue negative living cells were counted in all 9 quadrants. Cell number / ml was calculated with the following equation:

Cell number / ml = mean cell count / quadrant x dilution factor x 10^4

4.2.2 Isolation of primary B and T lymphocytes from different organs

For blood cells 20-30 µl blood was collected from the tail vein. For organ preparation mice were sacrificed with CO_2 and cranial dislocation. To get peritoneal lavages the peritoneal cavity was infused with 2 ml R10 medium, mice were massaged and a minunim of 1 ml with peritoneal cells was collected. Thymus was extracted after opening of the thorax and brought to single cell suspension through a cell strainer (70 µm, BD). Femora and tibia were opened on both sides and rinsed thoroughly with 5 ml R10 medium several times. Spleen and inguinal lymph nodes were collected after opening of the abdominal cavity on the side and brought to single cell suspension via a cell strainer (70 µm, BD). Cells were centrifuged at 1800 rpm (Heraeus Megafuge) for 8 min at 4°C.

4.2.2.1 Erythrocyte depletion

Single cell suspensions from different organs were resuspended in 5 ml erythrocyte lysis buffer (0.15 M NH_4Cl, 20 mM HEPES) after centrifugation and incubated for 5-7 min at room temperature. Lysis was stopped with 5 ml of the appropriate medium and centrifugation at 1800rpm (Heraeus Megafuge) for 8 min at 4°C. Cell pellets were resuspended in appropriate volumina for cell counting.

4.2.2.2 MACS-sorting of CD43⁻ splenic and CD19⁺ bone marrow B cells

Untouched splenic B cells were isolated by magnetic cell sorting (MACS) using the CD43 depletion method (Miltenyi Biotec). Bone marrow cells were sorted with CD19 magnetic beads. 1 x 10^7 cells were incubated with 10 µl magnetic beads (CD43 or CD19) and 90 µl MACS buffer (PBS, 2% FCS) for 20-30 min on ice. Cells were washed with 5 ml MACS buffer and resuspended in 1 ml MACS buffer after centrifugation for 5 min at 1800 rpm and 4°C.

After filtration with a Pre-Separation filter (Milteny Biotec) cells were sorted with an Auto-MACS (Miltenyi Biotec). CD19 stained bone marrow cells were sorted positively and CD43 stained splenic B cells were sorted negatively. Cells were suspended in R10 medium and kept on ice for further analyses. Purity was determined by flow cytometry.

4.2.3 Flow cytometry

4.2.3.1 Analyses of surface expression and intracellular proteins

1×10^5 - 2×10^6 cells were centrifuged at 1800 rpm and 4°C für for 8 min and washed with 1 ml FACS buffer (PBS, 2% FCS, 0.05% sodium azide). Surface staining was performed in 100 µl of the appropriate antibody combinations for 20 min on ice in the dark. Biotin-labeled antibodies were detected in a second step by appropriate secondary fluorochrome-conjugated streptavidin. Intracellular protein staining was performed according to the manufacturer's protocol of the Fix & Perm Kit (ADG Biosearch GmbH). After fixation and permeabilization cells were incubated in 20 µl of the appropriate antibody combination (6 fold concentrated) in FACS buffer for 20 min on ice. At the end or inbetween stainings probes were washed with 1 ml FACS buffer. Before analyses at the FACSCalibur flow cytometer (BD Bioscineces) cells were suspended in 300 - 400 µl FACS buffer and kept on ice. Flow cytometric data were analyzed using CellQUEST Pro software v4.0.2 (BD Biosciences).

4.2.3.2 MoFlo™ cell sorting

Cells were stained in 1 ml of the appropriate antibody combinations for 30 min on ice. Agglomerated cells were separated with a *Pre-Separation*-Filter (Miltenyi Biotec) and EDTA was added in a final concentration of 1 mM. Cell sorting was performed at a MoFlo™ cytometer (Dako Cytomation). Afterwards cells were collected and washed in the appropriate cell culture medium and kept on ice for further analyses such as cultivation or DNA/RNA preparation.

4.2.4 Histology

Spleens were frozen in Tissue Tek medium (Sakura) on dry ice and kept on -70°C. Cryosections (8 µm) were prepared at a Cryotom (CM3050S, Leica) with -15°C object temperature and -18°C chamber temperature, fixed in acetone for 10 min at -20°C, air dried and stored at -20°C. Sections were unfreezed for 15 min at room temperature, rehydrated for 5 min in PBS, blocked (PBS, 10% FCS, 0.1% BSA) for 30 min and stained with appropriate antibody dilutions in staining solution (PBS, 2% FCS, 0.05% Tween 20) for 30 min at room temperature in a dark wet chamber. Spare antibodies were removed by 2 washing steps for 10 min in wash buffer (PBS, 0.05% Tween 20) and a final washing step for 10 min in PBS. Sections were covered with 50 µl Moviol and hardened at 4°C. Stained sections were analyzed using an Axioplan 2 microscope (Carl Zeiss) equipped with Axiovision Rel 4.6 (Carl Zeiss) software.

4.2.5 Molecular biological methods

4.2.5.1 Plasmid DNA and RNA preparation

Isolation of plasmid DNA was performed using the commercially available Mini-, Midi- and Maxi-Kits according to manufacturer's instructions (Qiagen). For total RNA preparation from primary cells or cell lines the commercially available RNeasy-Kit (Qiagen) was used according to manufacturer's instructions. Isolation of DNA fragments from agarose gels was performed with the „QIAquick"-gel extraction kit (Qiagen) according to the manufacturer's protocol.

4.2.5.2 DNA restriction analyses

For digests with commercially available restriction enzymes (New England Biolabs) 0.02 µg/ml of isolated DNA were used with a specific enzyme activity of 1 – 2 U/µg DNA and addition of 1 x BSA. For analytical assays digests were performed at 37°C for 1 hour and for preparative digests time was extended to 2 hours.

4.2.5.3 DNA fragment agarose gel electrophoresis

DNA restriction fragments or PCR fragments with a size of 0.5 to 10 kbp were separated in a flat bed gel electrophoresis in 1% agarose (Peqlab). Smaller fragments were separated in 1.5% agarose. Agarose was boiled in the appropriate volume of 1 x TAE running buffer (50 x TAE-Laufpuffer: 2 M Tris-Cl, 1 M sodium acetate, 50 mM EDTA, pH mit HCl auf 8.5) to a clear solution. After cooling to 50°C, 5 µl ethidium bromide (10 mg/ml) were added per 100 ml 1 x TAE. Probes were loaded in the solidified gel and separated at a voltage of 100 V for 30 min.

4.2.5.4 Isolation of genomic tail DNA

Mouse tails were cut and transferred in 100-300 µl PBND buffer (50 KCl, 10 mM Tris-HCl pH 8.3, 2.5 mM $MgCl_2$-$6H_2O$, 0.1 mg/ml gelatine, 0.45% NP40, 0.45% Tween20) with 3 µl Proteinase K (Peqlab 20 mg/ml). Incubation was performed overnight at 56°C in a Thermo shaker. After Proteinase K heat inactivation at 95°C for 5 min probes were centriguged for 5 min at 13000 rpm (Heraeus Biofuge fresco). Supernatant was transferred into a new 1.5 ml reaction tube and kept on 4°C. For genotyping 3 µl total tail DNA was used.

4.2.5.5 DNA/RNA concentration

DNA/RNA concentrations were determined at an absorption at 260 nm and 280 nm using a NanoDrop (Peqlab).

4.2.5.6 Polymerase chain reaction (PCR)

Reactions were carried out in a final volume of 20 µl in a Themo Cycler from Applied Biosystems. DNA was denatured at 95°C for 2 min at the beginning and the cycles were repeated 30-35 times (95°C, 30 sec; 56-62°C, 30 sec ; 72°C, 60 sec). In the end fragments were elongated in a final step at 72°C for 4 min and probes were stored at 4°C. Annealing temperature was calculated by primer length and AT/GC distribution. Sample composition is shown in table 11 and assays were carried out in a Thermocycler GeneAmp 9700 (Applied Biosystems) after the following protocol.

Table 9. PCR component overwiew
Abbreviations: rev: *reverse*, fwd: *forward*, x: DNA volume depending on concentration.

components	volume
DNA (10-100 ng)	x µl
Primer_fwd (5µM)	1 µl
Primer_rev (5µM)	1 µl
dNTP-Mix (10mM)	0.5 µl
Taq-Puffer (10x Genaxxon)	2 µl
Taq-Polymerase	0,2 µl
dH_2O	ad 20 µl

4.2.5.7 cDNA synthesis

500 ng total RNA were reverse transcribed using the Fermentas cDNA-Kit (First Strand cDNA Synthesis Kit) with Oligo-dT12-18 primers according to the manufacturer's instructions (Fermentas).

4.2.5.8 Qualitative RT-PCR

For semi quantitative RT-PCR equal amounts of RNA were first reverse transcribed as described in 7.2.5.7. cDNA specific primer pairs were used for target gene amplification. Signal intensity on an agarose gel should be proportional to amount of specific mRNA in total RNA. For the control of loading and integrity of RNA HPRT cDNA amplification was carried out. Sample composition is shown in table 9 and assays were carried out in a Thermocycler GeneAmp 9700 (Applied Biosystems) after the following protocol.

1) initial denaturation:	94°C	3 min
2) denaturation:	94°C	20 s
3) annealing:	X°C	20 s
4) elongation:	72°C	25 s
5) final elongation:	72°C	4 min
6) steps 2-4	25-30 times	
6) storage:	4°C	∞

4.2.5.9 Quantitative TaqMan®-RT-PCR

For TaqMan PCR, total RNA was isolated from sorted B lymphoid populations with TRIZOL-Reagent according to the manufacturer's protocol (Invitrogen). RNA was cleaned up using the RNeasy-Kit (Qiagen). Reverse transcription was performed using the TaqMan® RNA-to-CT™ 1-Step Kit from ABI (Applied Biosystems). Real Time PCR was performed using 2 ng of reverse transcribed cDNA and commercially available TaqMan probes (ABI) according to the manufacturer's protocol (10 µl TaqMan Universal PCR Mix, 1 µl 20 x Taqman Gene expression assay, H_2O ad 20 µl). TaqMan PCR was carried out in triplicates with non-tempalte controls with a standard program (initial 10 min 95°C, followed by 40 cycles with 95°C 15 sec and 60°C 1 min) using an Applied Biosystems 7300 Re al-Time PCR system. Relative expression was calculated with the *comparative* CT method and fold expression was calculated against hypoxanthine-guanine phosphoribosyltransferase (HPRT).

4.2.5.10 Affymetrix microarray analysis

Total RNA was isolated from sorted B lymphoid populations with the TRIZOL-Reagent and purified with the RNeasy-Kit (Qiagen). RNA was processed as described (Wittmann et al., 2006). Microarray analyses using MOE430_2.0 *Affymetrix* DNA Chips were performed at the BioChip Facility of the University of Essen (Dr. L. Klein-Hitpass, Essen, Germany).

4.2.6 Antibody isotype detection using ELISA

Blood was taken at indicated time points and stored at -20°C. Total as well as antigen-specific Ig serum concentrations were determined by ELISA. Microtiter plates were coated overnight with goat anti-mouse IgM/IgG/IgA antibodies (Southern Biotechnology) for total Ig titers or with 1 µg/ml TNP-BSA (Biosearch Technologies) for antigen-specific titers at 4°C and blocked with PBS, 2% FCS for 30 min at room temperature. Sera were diluted in PBS, 2% FCS and incubated for 1 hour at room temperature. Isotype-specific HRP-conjugated detection antibodies were obtained from Southern Biotechnologies and used in a dilution of 1:2000. Color intensities are measured at 492 nm using a Spectramax 190 (Molecular Devices). Total Ig concentrations were calculated using isotype-specific standard curves, and antigen-specific Ig titers were calculated as corrected ODs using a defined standard and plate correction factors with the following equation: OD x dilution / (OD at defined standard concentration of plate 1 / OD at defined standard concentration of plate x).

4.2.7 IgG specific Elispot

Mice were sacrificed 14 days after boost immunization with TNP-KLH (100 µg). Cells from bone marrow and splenic cell suspensions were seeded in a concentration of 0.5×10^6 and 0.25×10^6 in triplicates on TNP-BSA (1 µg/ml) coated ELISpot plates and treated according to the manufacturer's protocol from Mabtech ELISpot^{PLUS} kit for mouse IgG (Mabtech). Spots were detected in an ELISpot reader using AID Elispot Version 2.0 software (Autoimmun Diagnostika GmbH).

4.2.8 Proteinbiochemistry

4.2.8.1 Cell lysates

Cells were washed in ice cold PBS and resuspended in 50 µl Net lysis buffer (50 mM Tris-HCl pH 7.5, 150 mM NaCl, 2 mM EDTA, 1% Triton-X-100, 10 µg/ml Aprotinin, 1 mM PMSF, 10 mM sodiumfluorid) per 1 x 10^7 cells. Lysis was performed for 20 min on ice, insoluble aggregates were centrifuged for 10 min at 4°C and 1 3000 rpm (Heraeus Biofuge fresco) and supernatant was transferred into a new 1.5 ml reaction tube and stored at -70°C.

4.2.8.2 Nuclear extracts

For cytosolic and nuclear extracts 1 x 10^7 (cell lines) or 5 x 10^7 (primary) cells were washed in ice cold PBS, resuspended in 100 µl ice cold cell lysis buffer (50 mM Tris-HCL pH 7.5, 25 mM NaCl, 1mM EDTA, 0.1% Triton X-100, 1mM PMSF) and incubated on ice for 30 min. Cells were decomposed with an cold homogenisator and centrifuged at 13000 rpm and 4°C for 20 min (Heraeus Biofuge fresco). Postnuclear supernatant was removed and stored at -70°C as cytosolic fraction. Pellet was resuspended in 10 µl ice cold nucleus lysis buffer (20 mM HEPES/KOH pH 7.9, 420 mM KCl, 1.5 mM $MgCl_2$, 0.1 mM EDTA, 20% glycerol, 4 mM PMSF), incubated on ice for 45 min and centrifuged for 30 min at 13000 rpm and 4°C. Supernatant contained nuclear fraction. Lysates were kept on -70°C for western blot analyses.

4.2.8.3 BCA test

For determination of cell lysate protein concentrations BCA protein assay (Thermo Scientific) was performed. Solution A and B were mixed in a ratio of 50+1 respectively. After incubation of 2 µl cell lysate with 200 µl of the mixture (A+B) for 15 min at 37°C absorption was measured at 595 nm in a Spectramax 190 (Molecular Devices). Concentration was calculated by a standard curve using defined BSA concentrations.

4.2.8.4 Phosphatase inhibitor and Ly294002 stimulation

For Pervanadat stimulation 100 µl of 20 mM orthovanadate were mixed with 230 µl of 30% H_2O_2 and incubated for 5 min at room temperature. Resulting 6 mM Pervanadat solution was diluted 1:6 in dH_2O to get a 1 mM Pervanadat stock solution for stimulation. Calyculin A

(Sigma Aldrich) and Okadaic acid potassium (Sigma Aldrich) were suspended in DMSO to a stock concentration of 100 µM and Ly294002 (Cell Signaling) was dissolved in DMSO to a stock concentration of 10 mM 1 x 10^6 cells were stimulated with 25 µM Pervanadat, 100 nM Calyculin A or 100 nM Okadaic acid in R10 medium for 10 or 20 min or with 5 and 50 µM Ly294002 for 10 and 20 min at 37°C and 5% CO_2 in an incubator. Stimulation was stopped with ice cold PBS and cell lysates were made as described in 7.2.8.1.

4.2.8.5 Immunoprecipitation

20 Mio NYC cells were washed twice with ice cold PBS and resuspended in 500 µl NET lysis buffer (50 mM Tris-HCl pH 7.5, 150 mM NaCl, 2 mM EDTA, 1% Triton-X-100, 10 µg/ml Aprotinin, 1 mM PMSF, 10 mM sodiumfluorid) containing 1mM sodium fluoride, 1mM Orthovanadate and 1mM PMSF (NET+). Lysis was performed for 20 min on ice followed by centrifugation for 10 min at 13000 rpm and 4°C (Her aeus Biofuge fresco). For preclearing of unspecific protein binding, supernatant was incubated with 30 µl prewashed protein-G-sepharose (Thermo scientific) for one hour at 4°C in an orbital shaker. After sedimentation for 5 sec at 13000 rpm and 4°C 10% of the supernatant (50 µl) were stored as loading control. For precipitation 3-5 µg of specific antibody were incubated with the remaining supernatant for two hours at 4°C on an orbital shaker. Afterwards lysate/antibody mixture was incubated with 50 µl prewashed protein-G-sepharose for another minimum of 3 hours at 4°C in an orbital shaker. Precipitates were washed 4 times with 500 µl NET+ buffer (centrifugation for 1 min at 5500 rpm and 4°C) and resuspended in 2 x SDS loadin g dye, boiled for 5 min at 95°C and used for western blot analyses.

4.2.8.6 SDS PAGE

Cell lysates were separated electrophoretically referring to molecular weight in an SDS polyacryamid gel (Laemmli, 1970). Electrophoresis was performed in a chamber from Hoefer. Probes were mixed with the appropriate volume of 5 x SDS loading buffer (250 mM Tris-HCl pH 6.8; 10% SDS, 50% glycerin, 5% bromphenol blue, 0.5 mM beta- mercaptoethanol), cooked at 95°C for 5 min and transferred onto the gel. Electrophoretic separation was performed at 40 mA per gel for 4 hours. For sizing of proteins a molecular weight standard (peqGold, Peqlab) was loaded next to the probes.

Table 10. Overview of components for SDS PAGE

components	resolving gel (10%) (ml)	stacking gel (ml)
Acrylamid/Bisacrylamid	7.5	0.875
0,5M Tris-Cl, pH 6,8	-	2,5
3M Tris-Cl, pH 8,8	3.75	-
dH_2O	18.135	6.525
10% SDS	0.3	0.1
10% APS	0.3	0.1
TEMED	0.015	0.006

4.2.8.7 Western blot analysis

Separated proteins were transferred onto a nitro cellulose membrane (Schleicher & Schuell) using semi-dry transfer in the *PerfectBlue*[TM]- semi-dry transfer system (Peqlab) according to the manufacturer's protocol. Proteins were stained with Ponceau-S (0.2% Ponceau S, 3% trichloroacetic acid, 3% sulfoalicyl acid) for protein loading. Membranes were blocked with 5% skimmed milk powder in TBS (10 x TBS: 200 mM Tris-Cl, pH 7.4; 1.5 M NaCl), 0.1% Tween 20 for 1 hour. Membranes were stained with appropriate primary followed by appropriate HRP-conjugated secondary antibodies for 1 hour. Inbetween membranes were washed 3 times for 10 min in TBS, 0.1% Tween 20. Signals were developed by the chemiluminescence method (Amersham Pharmacia). For further analyses membranes were stripped 2 times for 10 min in stripping buffer (0.1 M glycine pH 2.5, 0.5 M NaCl, 0.1% Tween 20, 0.1% NaN_3) and washed 3 times for 10 min in TBS, 0.1% Tween 20.

4.2.9 Calcium measurements

5×10^6 splenic cells were harvested and resuspended in 700 µl R10 medium with antibodies in defined concentrations and 1% Indo mix (Indo-1-AM (Molecular Probes), 458 µl DMSO 0.015% Pluronic F-127 (Molecular Probes) and incubated for 25 min at 30°C. After addition of 700 µl R 10 medium cells were further incubated for 10 min at 37°C. Cells were washed twice in 1 ml Krebs Ringer solution (10 mM HEPES pH 7,0, 140 mM NaCl, 4 mM KCl, 1 mM $MgCl_2$, 1 mM $CaCl_2$, 10 mM Glucose) and at the end resuspended in 500 µl Krebs Ringer solution.

Measurement was performed of 60 µl cell suspension in 540 µl warmed Krebs Ringer solution at a LSRII cytometer (BD Biosciences) using FACS DIVA software (BD Biosciences). Protocol included 40 sec baseline measurement followed by addition of 5 µg goat-anti-mouse IgM (µ-chain specific) Fab 2 fragments (Jackson Laboratories) for BCR stimulation for additional 3.5 min. Data were analyzed using FlowJo software v 8.8.6 (Tree Star, Inc.).

4.2.10 Transfection and infection of vertebrate cells

4.2.10.1 Transient transfection of adherent cells using calcium phosphate

One day prior to transfection 4×10^6 Phoenix-eco cells were seeded in a 10 cm well plate and incubated overnight at 37°C and 7.5% CO_2 in D10 medium. 20 µg of DNA were mixed with 125 µl 2 M $CaCl_2$ to a final volume of 1 ml with dH_2O. Under bubbling conditions mixture was added to 1 ml 2 x HBS (50 mM Hepes, 10 mM potassium chloride, 12 mM Dextrose, 280 mM sodium chloride, 1.5 mM Na_2HPO_4, pH 7.04) and incubated for 5 min at room temperature for precipitate formation. 2 ml medium was removed from seeded cells and 50 µl of 25 mM chloroquine were added. Precipitates were trickled onto cells and after 6-8 hours incubation at 37°C and 7.5% CO_2 cells were provided with fresh D10 medium. Viral supernatant is removed after 24, 48 and 72 h, filtered and stored at -70°C for infection.

4.2.10.2 Infection of NIH3T3 cells

25.000 NIH3T3 cells were seeded in a 24 well plate overnight at 37°C and 7.5% CO_2 in 1 ml D10 medium. Medium was removed and 900 µl of viral supernatant with 100 µl fresh D10 medium and 1 µl Polybrene were added. Cells were incubated 24 to 48 h at 37°C and 7.5% CO_2 before flow cytometric or western blot analyses.

4.2.11 *In vivo* methods

4.2.11.1 Immunization

For thymus-independent immune responses mice were challenged with 50 µg TNP-LPS (Biosearch Technologies) in PBS or 25 µg TNP-ficoll (Biosearch Technologies) in PBS intravenously. For thymus-dependent immune responses mice were injected with 100 µg TNP-KLH (Biosearch Technologies) in PBS + 100 µl complete Freud's adjuvants (CFA) intraperitoneally for primary immunization and with 100 µg TNP-KLH in PBS + 100 µl incomplete CFA (iCFA) intraperitoneally for boost immunization. Serum samples were taken at different time points and analyzed by antigen specific ELISA as described in 7.2.6.

4.2.11.2 FTY720-treatment

For FTY720 treatment 6-10 weeks old animals were injected intraperitoneally with either 20 μg of FTY720 (Cayman Chemicals) or an equivalent volume of saline. 4 hours after injection, animals were sacrificed; spleens were removed and processed for flow cytometry as described in 7.2.3.1 and cryosections as described in 7.2.4.

4.2.11.3 BrdU-treatment

For *in vivo* proliferation analyses mice were injected once with 2 mg BrdU (BD Pharmingen) intraperitoneally. Mice were sacrificed after 20 hours, organs prepared as described in 7.2.2 and cell populations analyzed using the BrdU Flow Kit Staining protocol (BD Biosciences) according to manufactorer's instructions by flow cytometry.

4.2.12 Statistics

All statistical analyses were performed using Graph-Pad Prism-Software (Graphpad software). Statistical significance was calculated by Mann-Whitney-U-test using two-tailed analysis with 95% confidence interval. Significance is shown as $p < 0.05$ *, $p < 0.01$ ** and $p < 0.001$ ***.

5 Epilogue

This work was supported by many people to whom I would like to express my gratitude. First of all I would like to thank Hans-Martin Jäck for ceding me this very interesting project, his support throughout the last years and very fruitful discussions. Falk Nimmerjahn and Martin Herrmann, I would like to thank for being members of my thesis advisory committee. Furthermore, I would like to thank Wolfgang Schuh and Lena Sandrock, the KLF2 group. They survived with me all ups and downs in the last three years. It was a really great time with you both and I had very much fun! Dirk Mielenz and Jürgen Wittmann, I would like to thank for advices and discussions. Martina Porstner, I would like to thank for all the nice times we spent together and especially for being my roommate on all our meetings. Edith Roth, I would like to thank for being my bench-neighbour and for all her advices and never ending good mood. Special thanks to all other members of the Jäck lab, the GK592 and the FOR832 for the unforgettable times we have spent together! Matthias Wabl and Gabriele Beck-Engeser, I would like to thank for the very nice and fruitful cooperation. It was really nice to work with you even with the big ocean lying between us. Above all I would like to thank my family and friends, especially my parents, grandparents and my sister, for their support and confidence. You have been there for me whenever I needed you! Last but not least I would like to thank my husband Thorsten for his patience, never ending optimism and faith in me!

6 References

Allende, M. L., Tuymetova, G., Lee, B. G., Bonifacino, E., Wu, Y. P., and Proia, R. L. (2010). S1P1 receptor directs the release of immature B cells from bone marrow into blood. J Exp Med 207, 1113-1124.

Allman, D., Lindsley, R. C., DeMuth, W., Rudd, K., Shinton, S. A., and Hardy, R. R. (2001). Resolution of three nonproliferative immature splenic B cell subsets reveals multiple selection points during peripheral B cell maturation. J Immunol 167, 6834-6840.

Allman, D., and Pillai, S. (2008). Peripheral B cell subsets. Curr Opin Immunol 20, 149-157.

Allman, D. M., Ferguson, S. E., and Cancro, M. P. (1992). Peripheral B cell maturation. I. Immature peripheral B cells in adults are heat-stable antigenhi and exhibit unique signaling characteristics. J Immunol 149, 2533-2540.

Allman, D. M., Ferguson, S. E., Lentz, V. M., and Cancro, M. P. (1993). Peripheral B cell maturation. II. Heat-stable antigen(hi) splenic B cells are an immature developmental intermediate in the production of long-lived marrow-derived B cells. J Immunol 151, 4431-4444.

Alt, F. W., Yancopoulos, G. D., Blackwell, T. K., Wood, C., Thomas, E., Boss, M., Coffman, R., Rosenberg, N., Tonegawa, S., and Baltimore, D. (1984). Ordered rearrangement of immunoglobulin heavy chain variable region segments. Embo J 3, 1209-1219.

Alugupalli, K. R., Gerstein, R. M., Chen, J., Szomolanyi-Tsuda, E., Woodland, R. T., and Leong, J. M. (2003). The resolution of relapsing fever borreliosis requires IgM and is concurrent with expansion of B1b lymphocytes. J Immunol 170, 3819-3827.

Alugupalli, K. R., Leong, J. M., Woodland, R. T., Muramatsu, M., Honjo, T., and Gerstein, R. M. (2004). B1b lymphocytes confer T cell-independent long-lasting immunity. Immunity 21, 379-390.

Amin, R. H., and Schlissel, M. S. (2008). Foxo1 directly regulates the transcription of recombination-activating genes during B cell development. Nat Immunol 9, 613-622.

Ansel, K. M., Harris, R. B., and Cyster, J. G. (2002). CXCL13 is required for B1 cell homing, natural antibody production, and body cavity immunity. Immunity 16, 67-76.

Ansel, K. M., Ngo, V. N., Hyman, P. L., Luther, S. A., Forster, R., Sedgwick, J. D., Browning, J. L., Lipp, M., and Cyster, J. G. (2000). A chemokine-driven positive feedback loop organizes lymphoid follicles. Nature 406, 309-314.

Attanavanich, K., and Kearney, J. F. (2004). Marginal zone, but not follicular B cells, are potent activators of naive CD4 T cells. J Immunol 172, 803-811.

Bai, A., Hu, H., Yeung, M., and Chen, J. (2007). Kruppel-like factor 2 controls T cell trafficking by activating L-selectin (CD62L) and sphingosine-1-phosphate receptor 1 transcription. J Immunol 178, 7632-7639.

Baumgarth, N., Herman, O. C., Jager, G. C., Brown, L. E., Herzenberg, L. A., and Chen, J. (2000). B-1 and B-2 cell-derived immunoglobulin M antibodies are nonredundant components of the protective response to influenza virus infection. J Exp Med 192, 271-280.

Berberich, S., Dahne, S., Schippers, A., Peters, T., Muller, W., Kremmer, E., Forster, R., and Pabst, O. (2008). Differential molecular and anatomical basis for B cell migration into the peritoneal cavity and omental milky spots. J Immunol 180, 2196-2203.

Berland, R., and Wortis, H. H. (2002). Origins and functions of B-1 cells with notes on the role of CD5. Annu Rev Immunol 20, 253-300.

Bhattacharya, D., Cheah, M. T., Franco, C. B., Hosen, N., Pin, C. L., Sha, W. C., and Weissman, I. L. (2007). Transcriptional profiling of antigen-dependent murine B cell differentiation and memory formation. J Immunol 179, 6808-6819.

Blom, N., Gammeltoft, S., and Brunak, S. (1999). Sequence and structure-based prediction of eukaryotic protein phosphorylation sites. J Mol Biol 294, 1351-1362.

Boes, M., Prodeus, A. P., Schmidt, T., Carroll, M. C., and Chen, J. (1998). A critical role of natural immunoglobulin M in immediate defense against systemic bacterial infection. J Exp Med 188, 2381-2386.

Briles, D. E., Forman, C., Hudak, S., and Claflin, J. L. (1982). Anti-phosphorylcholine antibodies of the T15 idiotype are optimally protective against Streptococcus pneumoniae. J Exp Med 156, 1177-1185.

Brouns, G. S., de Vries, E., van Noesel, C. J., Mason, D. Y., van Lier, R. A., and Borst, J. (1993). The structure of the mu/pseudo light chain complex on human pre-B cells is consistent with a function in signal transduction. Eur J Immunol 23, 1088-1097.

Brownawell, A. M., Kops, G. J., Macara, I. G., and Burgering, B. M. (2001). Inhibition of nuclear import by protein kinase B (Akt) regulates the subcellular distribution and activity of the forkhead transcription factor AFX. Mol Cell Biol 21, 3534-3546.

Brunet, A., Bonni, A., Zigmond, M. J., Lin, M. Z., Juo, P., Hu, L. S., Anderson, M. J., Arden, K. C., Blenis, J., and Greenberg, M. E. (1999). Akt promotes cell survival by phosphorylating and inhibiting a Forkhead transcription factor. Cell 96, 857-868.

Buckley, A. F., Kuo, C. T., and Leiden, J. M. (2001). Transcription factor LKLF is sufficient to program T cell quiescence via a c-Myc--dependent pathway. Nat Immunol 2, 698-704.

Burgering, B. M., and Kops, G. J. (2002). Cell cycle and death control: long live Forkheads. Trends Biochem Sci 27, 352-360.

Cariappa, A., Tang, M., Parng, C., Nebelitskiy, E., Carroll, M., Georgopoulos, K., and Pillai, S. (2001). The follicular versus marginal zone B lymphocyte cell fate decision is regulated by Aiolos, Btk, and CD21. Immunity 14, 603-615.

Carlson, C. M., Endrizzi, B. T., Wu, J., Ding, X., Weinreich, M. A., Walsh, E. R., Wani, M. A., Lingrel, J. B., Hogquist, K. A., and Jameson, S. C. (2006). Kruppel-like factor 2 regulates thymocyte and T-cell migration. Nature 442, 299-302.

Cassese, G., Arce, S., Hauser, A. E., Lehnert, K., Moewes, B., Mostarac, M., Muehlinghaus, G., Szyska, M., Radbruch, A., and Manz, R. A. (2003). Plasma cell survival is mediated by synergistic effects of cytokines and adhesion-dependent signals. J Immunol 171, 1684-1690.

Cattoretti, G., Chang, C. C., Cechova, K., Zhang, J., Ye, B. H., Falini, B., Louie, D. C., Offit, K., Chaganti, R. S., and Dalla-Favera, R. (1995). BCL-6 protein is expressed in germinal-center B cells. Blood 86, 45-53.

Chen, J., Limon, J. J., Blanc, C., Peng, S. L., and Fruman, D. A. (2010). Foxo1 regulates marginal zone B-cell development. Eur J Immunol 40, 1890-1896.

Cinamon, G., Matloubian, M., Lesneski, M. J., Xu, Y., Low, C., Lu, T., Proia, R. L., and Cyster, J. G. (2004). Sphingosine 1-phosphate receptor 1 promotes B cell localization in the splenic marginal zone. Nat Immunol 5, 713-720.

Cinamon, G., Zachariah, M. A., Lam, O. M., Foss, F. W., Jr., and Cyster, J. G. (2008). Follicular shuttling of marginal zone B cells facilitates antigen transport. Nat Immunol 9, 54-62.

Clayton, E., Bardi, G., Bell, S. E., Chantry, D., Downes, C. P., Gray, A., Humphries, L. A., Rawlings, D., Reynolds, H., Vigorito, E., and Turner, M. (2002). A crucial role for the p110delta subunit of phosphatidylinositol 3-kinase in B cell development and activation. J Exp Med 196, 753-763.

Conkright, M. D., Wani, M. A., and Lingrel, J. B. (2001). Lung Kruppel-like factor contains an autoinhibitory domain that regulates its transcriptional activation by binding WWP1, an E3 ubiquitin ligase. J Biol Chem 276, 29299-29306.

Cyster, J. G. (2000). B cells on the front line. Nat Immunol 1, 9-10.

Cyster, J. G. (2003). Homing of antibody secreting cells. Immunol Rev 194, 48-60.

Das, H., Kumar, A., Lin, Z., Patino, W. D., Hwang, P. M., Feinberg, M. W., Majumder, P. K., and Jain, M. K. (2006). Kruppel-like factor 2 (KLF2) regulates proinflammatory activation of monocytes. Proc Natl Acad Sci U S A 103, 6653-6658.

Deane, J. A., and Fruman, D. A. (2004). Phosphoinositide 3-kinase: diverse roles in immune cell activation. Annu Rev Immunol 22, 563-598.

Dengler, H. S., Baracho, G. V., Omori, S. A., Bruckner, K., Arden, K. C., Castrillon, D. H., DePinho, R. A., and Rickert, R. C. (2008). Distinct functions for the transcription factor Foxo1 at various stages of B cell differentiation. Nat Immunol 9, 1388-1398.

Dent, A. L., Shaffer, A. L., Yu, X., Allman, D., and Staudt, L. M. (1997). Control of inflammation, cytokine expression, and germinal center formation by BCL-6. Science 276, 589-592.

Dießenbacher, P. (2005). Prozessierung des Zinkfinger-Transkriptionsfaktors LKLF im Verlauf der Aktivierung von T-Zellen, PhD thesis. Institut für Immunologie des Universitätsklinikums Hamburg-Eppendorf.

Dorshkind, K., and Montecino-Rodriguez, E. (2007). Fetal B-cell lymphopoiesis and the emergence of B-1-cell potential. Nat Rev Immunol 7, 213-219.

Edry, E., and Melamed, D. (2004). Receptor editing in positive and negative selection of B lymphopoiesis. J Immunol 173, 4265-4271.

Fabre, S., Carrette, F., Chen, J., Lang, V., Semichon, M., Denoyelle, C., Lazar, V., Cagnard, N., Dubart-Kupperschmitt, A., Mangeney, M., et al. (2008). FOXO1 regulates L-Selectin and a network of human T cell homing molecules downstream of phosphatidylinositol 3-kinase. J Immunol 181, 2980-2989.

Fang, W., Mueller, D. L., Pennell, C. A., Rivard, J. J., Li, Y. S., Hardy, R. R., Schlissel, M. S., and Behrens, T. W. (1996). Frequent aberrant immunoglobulin gene rearrangements in pro-B cells revealed by a bcl-xL transgene. Immunity 4, 291-299.

Faust, E. A., Saffran, D. C., Toksoz, D., Williams, D. A., and Witte, O. N. (1993). Distinctive growth requirements and gene expression patterns distinguish progenitor B cells from pre-B cells. J Exp Med 177, 915-923.

Fruman, D. A., Ferl, G. Z., An, S. S., Donahue, A. C., Satterthwaite, A. B., and Witte, O. N. (2002). Phosphoinositide 3-kinase and Bruton's tyrosine kinase regulate overlapping sets of genes in B lymphocytes. Proc Natl Acad Sci U S A 99, 359-364.

Fukuda, T., Yoshida, T., Okada, S., Hatano, M., Miki, T., Ishibashi, K., Okabe, S., Koseki, H., Hirosawa, S., Taniguchi, M., et al. (1997). Disruption of the Bcl6 gene results in an impaired germinal center formation. J Exp Med 186, 439-448.

Gay, D., Saunders, T., Camper, S., and Weigert, M. (1993). Receptor editing: an approach by autoreactive B cells to escape tolerance. J Exp Med 177, 999-1008.

Geier, J. K., and Schlissel, M. S. (2006). Pre-BCR signals and the control of Ig gene rearrangements. Semin Immunol 18, 31-39.

Glynne, R., Ghandour, G., Rayner, J., Mack, D. H., and Goodnow, C. C. (2000). B-lymphocyte quiescence, tolerance and activation as viewed by global gene expression profiling on microarrays. Immunol Rev 176, 216-246.

Grawunder, U., Leu, T. M., Schatz, D. G., Werner, A., Rolink, A. G., Melchers, F., and Winkler, T. H. (1995). Down-regulation of RAG1 and RAG2 gene expression in preB cells after functional immunoglobulin heavy chain rearrangement. Immunity 3, 601-608.

Grayson, J. M., Murali-Krishna, K., Altman, J. D., and Ahmed, R. (2001). Gene expression in antigen-specific CD8+ T cells during viral infection. J Immunol 166, 795-799.

Gubbels Bupp, M. R., Edwards, B., Guo, C., Wei, D., Chen, G., Wong, B., Masterler, E., and Peng, S. L. (2009). T cells require Foxo1 to populate the peripheral lymphoid organs. Eur J Immunol 39, 2991-2999.

Gutman, G. A., Warner, N. L., and Harris, A. W. (1981). Immunoglobulin production by murine B-lymphoma cells. Clin Immunol Immunopathol 18, 230-244.

Ha, S. A., Tsuji, M., Suzuki, K., Meek, B., Yasuda, N., Kaisho, T., and Fagarasan, S. (2006). Regulation of B1 cell migration by signals through Toll-like receptors. J Exp Med 203, 2541-2550.

Haaland, R. E., Yu, W., and Rice, A. P. (2005). Identification of LKLF-regulated genes in quiescent CD4+ T lymphocytes. Mol Immunol 42, 627-641.

Hamann, A., Andrew, D. P., Jablonski-Westrich, D., Holzmann, B., and Butcher, E. C. (1994). Role of alpha 4-integrins in lymphocyte homing to mucosal tissues in vivo. J Immunol 152, 3282-3293.

Hargreaves, D. C., Hyman, P. L., Lu, T. T., Ngo, V. N., Bidgol, A., Suzuki, G., Zou, Y. R., Littman, D. R., and Cyster, J. G. (2001). A coordinated change in chemokine responsiveness guides plasma cell movements. J Exp Med 194, 45-56.

Hart, G. T., Wang, X., Hogquist, K. A., and Jameson, S. C. (2011). Kruppel-like factor 2 (KLF2) regulates B-cell reactivity, subset differentiation, and trafficking molecule expression. Proc Natl Acad Sci U S A 108, 716-721.

Hatakeyama, S., and Nakayama, K. I. (2003). U-box proteins as a new family of ubiquitin ligases. Biochem Biophys Res Commun *302*, 635-645.

Haughton, G., Arnold, L. W., Bishop, G. A., and Mercolino, T. J. (1986). The CH series of murine B cell lymphomas: neoplastic analogues of Ly-1+ normal B cells. Immunol Rev *93*, 35-51.

Hauser, A. E., Debes, G. F., Arce, S., Cassese, G., Hamann, A., Radbruch, A., and Manz, R. A. (2002). Chemotactic responsiveness toward ligands for CXCR3 and CXCR4 is regulated on plasma blasts during the time course of a memory immune response. J Immunol *169*, 1277-1282.

Hauser, A. E., Muehlinghaus, G., Manz, R. A., Cassese, G., Arce, S., Debes, G. F., Hamann, A., Berek, C., Lindenau, S., Doerner, T., et al. (2003). Long-lived plasma cells in immunity and inflammation. Ann N Y Acad Sci *987*, 266-269.

Herzog, S., Hug, E., Meixlsperger, S., Paik, J. H., DePinho, R. A., Reth, M., and Jumaa, H. (2008). SLP-65 regulates immunoglobulin light chain gene recombination through the PI(3)K-PKB-Foxo pathway. Nat Immunol *9*, 623-631.

Herzog, S., Reth, M., and Jumaa, H. (2009). Regulation of B-cell proliferation and differentiation by pre-B-cell receptor signalling. Nat Rev Immunol *9*, 195-205.

Hess, J., Werner, A., Wirth, T., Melchers, F., Jäck, H. M., and Winkler, T. H. (2001). Induction of pre-B cell proliferation after de novo synthesis of the pre-B cell receptor. Proc Natl Acad Sci U S A *98*, 1745-1750.

Hobeika, E., Thiemann, S., Storch, B., Jumaa, H., Nielsen, P. J., Pelanda, R., and Reth, M. (2006). Testing gene function early in the B cell lineage in mb1-cre mice. Proc Natl Acad Sci U S A *103*, 13789-13794.

Hoek, K. L., Gordy, L. E., Collins, P. L., Parekh, V. V., Aune, T. M., Joyce, S., Thomas, J. W., Van Kaer, L., and Sebzda, E. (2010). Follicular B Cell Trafficking within the Spleen Actively Restricts Humoral Immune Responses. Immunity *33*, 254-265.

Holodick, N. E., Tumang, J. R., and Rothstein, T. L. (2010). Immunoglobulin secretion by B1 cells: Differential intensity and IRF4-Dependence of Spontaneous IgM secretion by peritoneal and splenic B1 cells. Eur J Immunology, Accepted manuscript online: 2 SEP 2010, DOI: 2010.1002/eji.201040545.

Hombach, J., Tsubata, T., Leclercq, L., Stappert, H., and Reth, M. (1990). Molecular components of the B-cell antigen receptor complex of the IgM class. Nature *343*, 760-762.

Honjo, T., Kinoshita, K., and Muramatsu, M. (2002). Molecular mechanism of class switch recombination: linkage with somatic hypermutation. Annu Rev Immunol *20*, 165-196.

Hozumi, K., Negishi, N., Suzuki, D., Abe, N., Sotomaru, Y., Tamaoki, N., Mailhos, C., Ish-Horowicz, D., Habu, S., and Owen, M. J. (2004). Delta-like 1 is necessary for the generation of marginal zone B cells but not T cells in vivo. Nat Immunol *5*, 638-644.

Huang, K., Johnson, K. D., Petcherski, A. G., Vandergon, T., Mosser, E. A., Copeland, N. G., Jenkins, N. A., Kimble, J., and Bresnick, E. H. (2000). A HECT domain ubiquitin ligase closely related to the mammalian protein WWP1 is essential for Caenorhabditis elegans embryogenesis. Gene *252*, 137-145.

Ishihara, H., Martin, B. L., Brautigan, D. L., Karaki, H., Ozaki, H., Kato, Y., Fusetani, N., Watabe, S., Hashimoto, K., Uemura, D., and et al. (1989). Calyculin A and okadaic acid: inhibitors of protein phosphatase activity. Biochem Biophys Res Commun *159*, 871-877.

Jacob, J., Kassir, R., and Kelsoe, G. (1991). In situ studies of the primary immune response to (4-hydroxy-3-nitrophenyl)acetyl. I. The architecture and dynamics of responding cell populations. J Exp Med *173*, 1165-1175.

Jainchill, J. L., Aaronson, S. A., and Todaro, G. J. (1969). Murine sarcoma and leukemia viruses: assay using clonal lines of contact-inhibited mouse cells. J Virol *4*, 549-553.

Jou, S. T., Carpino, N., Takahashi, Y., Piekorz, R., Chao, J. R., Carpino, N., Wang, D., and Ihle, J. N. (2002). Essential, nonredundant role for the phosphoinositide 3-kinase p110delta in signaling by the B-cell receptor complex. Mol Cell Biol *22*, 8580-8591.

Jung, D., Giallourakis, C., Mostoslavsky, R., and Alt, F. W. (2006). Mechanism and control of V(D)J recombination at the immunoglobulin heavy chain locus. Annu Rev Immunol *24*, 541-570.

Kabashima, K., Haynes, N. M., Xu, Y., Nutt, S. L., Allende, M. L., Proia, R. L., and Cyster, J. G. (2006). Plasma cell S1P1 expression determines secondary lymphoid organ retention versus bone marrow tropism. J Exp Med 203, 2683-2690.
Kaczynski, J., Cook, T., and Urrutia, R. (2003). Sp1- and Kruppel-like transcription factors. Genome Biol 4, 206.
Karasuyama, H., Kudo, A., and Melchers, F. (1990). The proteins encoded by the VpreB and lambda 5 pre-B cell-specific genes can associate with each other and with mu heavy chain. J Exp Med 172, 969-972.
Keyna, U., Applequist, S. E., Jongstra, J., Beck-Engeser, G. B., and Jäck, H. M. (1995). Ig mu heavy chains with VH81X variable regions do not associate with lambda 5. Ann N Y Acad Sci 764, 39-42.
Kharas, M. G., Yusuf, I., Scarfone, V. M., Yang, V. W., Segre, J. A., Huettner, C. S., and Fruman, D. A. (2007). KLF4 suppresses transformation of pre-B cells by ABL oncogenes. Blood 109, 747-755.
Klaewsongkram, J., Yang, Y., Golech, S., Katz, J., Kaestner, K. H., and Weng, N. P. (2007). Kruppel-like factor 4 regulates B cell number and activation-induced B cell proliferation. J Immunol 179, 4679-4684.
Kline, G. H., Hartwell, L., Beck-Engeser, G. B., Keyna, U., Zaharevitz, S., Klinman, N. R., and Jäck, H. M. (1998). Pre-B cell receptor-mediated selection of pre-B cells synthesizing functional mu heavy chains. J Immunol 161, 1608-1618.
Kunisawa, J., Kurashima, Y., Gohda, M., Higuchi, M., Ishikawa, I., Miura, F., Ogahara, I., and Kiyono, H. (2007). Sphingosine 1-phosphate regulates peritoneal B-cell trafficking for subsequent intestinal IgA production. Blood 109, 3749-3756.
Kunkel, E. J., and Butcher, E. C. (2003). Plasma-cell homing. Nat Rev Immunol 3, 822-829.
Kuo, C. T., Veselits, M. L., and Leiden, J. M. (1997). LKLF: A transcriptional regulator of single-positive T cell quiescence and survival. Science 277, 1986-1990.
Kuroda, K., Han, H., Tani, S., Tanigaki, K., Tun, T., Furukawa, T., Taniguchi, Y., Kurooka, H., Hamada, Y., Toyokuni, S., and Honjo, T. (2003). Regulation of marginal zone B cell development by MINT, a suppressor of Notch/RBP-J signaling pathway. Immunity 18, 301-312.
Langerak, A. W., and van Dongen, J. J. (2006). Recombination in the human IGK locus. Crit Rev Immunol 26, 23-42.
Leadbetter, E. A., Brigl, M., Illarionov, P., Cohen, N., Luteran, M. C., Pillai, S., Besra, G. S., and Brenner, M. B. (2008). NK T cells provide lipid antigen-specific cognate help for B cells. Proc Natl Acad Sci U S A 105, 8339-8344.
Lee, J. S., Yu, Q., Shin, J. T., Sebzda, E., Bertozzi, C., Chen, M., Mericko, P., Stadtfeld, M., Zhou, D., Cheng, L., et al. (2006). Klf2 is an essential regulator of vascular hemodynamic forces in vivo. Dev Cell 11, 845-857.
Loder, F., Mutschler, B., Ray, R. J., Paige, C. J., Sideras, P., Torres, R., Lamers, M. C., and Carsetti, R. (1999). B cell development in the spleen takes place in discrete steps and is determined by the quality of B cell receptor-derived signals. J Exp Med 190, 75-89.
Lopes-Carvalho, T., and Kearney, J. F. (2004). Development and selection of marginal zone B cells. Immunol Rev 197, 192-205.
Lu, L., and Osmond, D. G. (1997). Apoptosis during B lymphopoiesis in mouse bone marrow. J Immunol 158, 5136-5145.
Ma, S., Pathak, S., Mandal, M., Trinh, L., Clark, M. R., and Lu, R. (2010). Ikaros and Aiolos inhibit pre-B-cell proliferation by directly suppressing c-Myc expression. Mol Cell Biol 30, 4149-4158.
Ma, S., Pathak, S., Trinh, L., and Lu, R. (2008). Interferon regulatory factors 4 and 8 induce the expression of Ikaros and Aiolos to down-regulate pre-B-cell receptor and promote cell-cycle withdrawal in pre-B-cell development. Blood 111, 1396-1403.
Makowska, A., Faizunnessa, N. N., Anderson, P., Midtvedt, T., and Cardell, S. (1999). CD1high B cells: a population of mixed origin. Eur J Immunol 29, 3285-3294.
Manning, B. D., and Cantley, L. C. (2007). AKT/PKB signaling: navigating downstream. Cell 129, 1261-1274.

Martin, F., and Kearney, J. F. (2000). Positive selection from newly formed to marginal zone B cells depends on the rate of clonal production, CD19, and btk. Immunity 12, 39-49.
Martin, F., and Kearney, J. F. (2002). Marginal-zone B cells. Nat Rev Immunol 2, 323-335.
Maruyama, M., Lam, K. P., and Rajewsky, K. (2000). Memory B-cell persistence is independent of persisting immunizing antigen. Nature 407, 636-642.
Matthias, P., and Rolink, A. G. (2005). Transcriptional networks in developing and mature B cells. Nat Rev Immunol 5, 497-508.
McGuire, K. L., and Vitetta, E. S. (1981). kappa/lambda Shifts do not occur during maturation of murine B cells. J Immunol 127, 1670-1673.
McHeyzer-Williams, L. J., Driver, D. J., and McHeyzer-Williams, M. G. (2001). Germinal center reaction. Curr Opin Hematol 8, 52-59.
McHeyzer-Williams, M. G. (2003). B cells as effectors. Curr Opin Immunol 15, 354-361.
McHeyzer-Williams, M. G., and Ahmed, R. (1999). B cell memory and the long-lived plasma cell. Curr Opin Immunol 11, 172-179.
Melchers, F., Rolink, A., Grawunder, U., Winkler, T. H., Karasuyama, H., Ghia, P., and Andersson, J. (1995). Positive and negative selection events during B lymphopoiesis. Curr Opin Immunol 7, 214-227.
Merrell, K. T., Benschop, R. J., Gauld, S. B., Aviszus, K., Decote-Ricardo, D., Wysocki, L. J., and Cambier, J. C. (2006). Identification of anergic B cells within a wild-type repertoire. Immunity 25, 953-962.
Minges Wols, H. A., Underhill, G. H., Kansas, G. S., and Witte, P. L. (2002). The role of bone marrow-derived stromal cells in the maintenance of plasma cell longevity. J Immunol 169, 4213-4221.
Mittrucker, H. W., Matsuyama, T., Grossman, A., Kundig, T. M., Potter, J., Shahinian, A., Wakeham, A., Patterson, B., Ohashi, P. S., and Mak, T. W. (1997). Requirement for the transcription factor LSIRF/IRF4 for mature B and T lymphocyte function. Science 275, 540-543.
Mombaerts, P., Iacomini, J., Johnson, R. S., Herrup, K., Tonegawa, S., and Papaioannou, V. E. (1992). RAG-1-deficient mice have no mature B and T lymphocytes. Cell 68, 869-877.
Mora, J. R., and von Andrian, U. H. (2008). Differentiation and homing of IgA-secreting cells. Mucosal Immunol 1, 96-109.
Moran, S. T., Cariappa, A., Liu, H., Muir, B., Sgroi, D., Boboila, C., and Pillai, S. (2007). Synergism between NF-kappa B1/p50 and Notch2 during the development of marginal zone B lymphocytes. J Immunol 179, 195-200.
Mosser, E. A., Kasanov, J. D., Forsberg, E. C., Kay, B. K., Ney, P. A., and Bresnick, E. H. (1998). Physical and functional interactions between the transactivation domain of the hematopoietic transcription factor NF-E2 and WW domains. Biochemistry 37, 13686-13695.
Mumby, M. C., and Walter, G. (1993). Protein serine/threonine phosphatases: structure, regulation, and functions in cell growth. Physiol Rev 73, 673-699.
Muramatsu, M., Kinoshita, K., Fagarasan, S., Yamada, S., Shinkai, Y., and Honjo, T. (2000). Class switch recombination and hypermutation require activation-induced cytidine deaminase (AID), a potential RNA editing enzyme. Cell 102, 553-563.
Muto, A., Tashiro, S., Nakajima, O., Hoshino, H., Takahashi, S., Sakoda, E., Ikebe, D., Yamamoto, M., and Igarashi, K. (2004). The transcriptional programme of antibody class switching involves the repressor Bach2. Nature 429, 566-571.
Nemazee, D., and Weigert, M. (2000). Revising B cell receptors. J Exp Med 191, 1813-1817.
Nie, Y., Waite, J., Brewer, F., Sunshine, M. J., Littman, D. R., and Zou, Y. R. (2004). The role of CXCR4 in maintaining peripheral B cell compartments and humoral immunity. J Exp Med 200, 1145-1156.
O'Connor, B. P., Raman, V. S., Erickson, L. D., Cook, W. J., Weaver, L. K., Ahonen, C., Lin, L. L., Mantchev, G. T., Bram, R. J., and Noelle, R. J. (2004). BCMA is essential for the survival of long-lived bone marrow plasma cells. J Exp Med 199, 91-98.
Ochsenbein, A. F., Fehr, T., Lutz, C., Suter, M., Brombacher, F., Hengartner, H., and Zinkernagel, R. M. (1999). Control of early viral and bacterial distribution and disease by natural antibodies. Science 286, 2156-2159.

Odendahl, M., Mei, H., Hoyer, B. F., Jacobi, A. M., Hansen, A., Muehlinghaus, G., Berek, C., Hiepe, F., Manz, R., Radbruch, A., and Dorner, T. (2005). Generation of migratory antigen-specific plasma blasts and mobilization of resident plasma cells in a secondary immune response. Blood 105, 1614-1621.

Okada, T., Ngo, V. N., Ekland, E. H., Forster, R., Lipp, M., Littman, D. R., and Cyster, J. G. (2002). Chemokine requirements for B cell entry to lymph nodes and Peyer's patches. J Exp Med 196, 65-75.

Okkenhaug, K., Bilancio, A., Farjot, G., Priddle, H., Sancho, S., Peskett, E., Pearce, W., Meek, S. E., Salpekar, A., Waterfield, M. D., et al. (2002). Impaired B and T cell antigen receptor signaling in p110delta PI 3-kinase mutant mice. Science 297, 1031-1034.

Okkenhaug, K., and Vanhaesebroeck, B. (2003). PI3K in lymphocyte development, differentiation and activation. Nat Rev Immunol 3, 317-330.

Oliver, A. M., Martin, F., Gartland, G. L., Carter, R. H., and Kearney, J. F. (1997). Marginal zone B cells exhibit unique activation, proliferative and immunoglobulin secretory responses. Eur J Immunol 27, 2366-2374.

Oliver, A. M., Martin, F., and Kearney, J. F. (1999). IgMhighCD21high lymphocytes enriched in the splenic marginal zone generate effector cells more rapidly than the bulk of follicular B cells. J Immunol 162, 7198-7207.

Omori, S. A., Cato, M. H., Anzelon-Mills, A., Puri, K. D., Shapiro-Shelef, M., Calame, K., and Rickert, R. C. (2006). Regulation of class-switch recombination and plasma cell differentiation by phosphatidylinositol 3-kinase signaling. Immunity 25, 545-557.

Osmond, D. G. (1991). Proliferation kinetics and the lifespan of B cells in central and peripheral lymphoid organs. Curr Opin Immunol 3, 179-185.

Osmond, D. G. (1993). The turnover of B-cell populations. Immunol Today 14, 34-37.

Pear, W. S., Nolan, G. P., Scott, M. L., and Baltimore, D. (1993). Production of high-titer helper-free retroviruses by transient transfection. Proc Natl Acad Sci U S A 90, 8392-8396.

Pearson, R., Fleetwood, J., Eaton, S., Crossley, M., and Bao, S. (2008). Kruppel-like transcription factors: a functional family. Int J Biochem Cell Biol 40, 1996-2001.

Pereira, J. P., Cyster, J. G., and Xu, Y. (2010). A role for S1P and S1P1 in immature-B cell egress from mouse bone marrow. PLoS One 5, e9277.

Radic, M. Z., Erikson, J., Litwin, S., and Weigert, M. (1993). B lymphocytes may escape tolerance by revising their antigen receptors. J Exp Med 177, 1165-1173.

Reimold, A. M., Iwakoshi, N. N., Manis, J., Vallabhajosyula, P., Szomolanyi-Tsuda, E., Gravallese, E. M., Friend, D., Grusby, M. J., Alt, F., and Glimcher, L. H. (2001). Plasma cell differentiation requires the transcription factor XBP-1. Nature 412, 300-307.

Rena, G., Prescott, A. R., Guo, S., Cohen, P., and Unterman, T. G. (2001). Roles of the forkhead in rhabdomyosarcoma (FKHR) phosphorylation sites in regulating 14-3-3 binding, transactivation and nuclear targetting. Biochem J 354, 605-612.

Rolink, A., Grawunder, U., Winkler, T. H., Karasuyama, H., and Melchers, F. (1994). IL-2 receptor alpha chain (CD25, TAC) expression defines a crucial stage in pre-B cell development. Int Immunol 6, 1257-1264.

Rolink, A. G., Winkler, T., Melchers, F., and Andersson, J. (2000). Precursor B cell receptor-dependent B cell proliferation and differentiation does not require the bone marrow or fetal liver environment. J Exp Med 191, 23-32.

Saito, T., Chiba, S., Ichikawa, M., Kunisato, A., Asai, T., Shimizu, K., Yamaguchi, T., Yamamoto, G., Seo, S., Kumano, K., et al. (2003). Notch2 is preferentially expressed in mature B cells and indispensable for marginal zone B lineage development. Immunity 18, 675-685.

Schmidt-Supprian, M., Wunderlich, F. T., and Rajewsky, K. (2007). Excision of the Frt-flanked neo (R) cassette from the CD19cre knock-in transgene reduces Cre-mediated recombination. Transgenic Res 16, 657-660.

Schober, S. L., Kuo, C. T., Schluns, K. S., Lefrancois, L., Leiden, J. M., and Jameson, S. C. (1999). Expression of the transcription factor lung Kruppel-like factor is regulated by cytokines and correlates with survival of memory T cells in vitro and in vivo. J Immunol 163, 3662-3667.

Schuh, W., Meister, S., Herrmann, K., Bradl, H., and Jäck, H. M. (2008). Transcriptome analysis in primary B lymphoid precursors following induction of the pre-B cell receptor. Mol Immunol 45, 362-375.
Sebzda, E., Zou, Z., Lee, J. S., Wang, T., and Kahn, M. L. (2008). Transcription factor KLF2 regulates the migration of naive T cells by restricting chemokine receptor expression patterns. Nat Immunol 9, 292-300.
Shaffer, A. L., Yu, X., He, Y., Boldrick, J., Chan, E. P., and Staudt, L. M. (2000). BCL-6 represses genes that function in lymphocyte differentiation, inflammation, and cell cycle control. Immunity 13, 199-212.
Shapiro-Shelef, M., and Calame, K. (2005). Regulation of plasma-cell development. Nat Rev Immunol 5, 230-242.
Shapiro-Shelef, M., Lin, K. I., McHeyzer-Williams, L. J., Liao, J., McHeyzer-Williams, M. G., and Calame, K. (2003). Blimp-1 is required for the formation of immunoglobulin secreting plasma cells and pre-plasma memory B cells. Immunity 19, 607-620.
Shinkai, Y., Rathbun, G., Lam, K. P., Oltz, E. M., Stewart, V., Mendelsohn, M., Charron, J., Datta, M., Young, F., Stall, A. M., and et al. (1992). RAG-2-deficient mice lack mature lymphocytes owing to inability to initiate V(D)J rearrangement. Cell 68, 855-867.
Sierro, F., Biben, C., Martinez-Munoz, L., Mellado, M., Ransohoff, R. M., Li, M., Woehl, B., Leung, H., Groom, J., Batten, M., et al. (2007). Disrupted cardiac development but normal hematopoiesis in mice deficient in the second CXCL12/SDF-1 receptor, CXCR7. Proc Natl Acad Sci U S A 104, 14759-14764.
Sinclair, L. V., Finlay, D., Feijoo, C., Cornish, G. H., Gray, A., Ager, A., Okkenhaug, K., Hagenbeek, T. J., Spits, H., and Cantrell, D. A. (2008). Phosphatidylinositol-3-OH kinase and nutrient-sensing mTOR pathways control T lymphocyte trafficking. Nat Immunol 9, 513-521.
Song, H., and Cerny, J. (2003). Functional heterogeneity of marginal zone B cells revealed by their ability to generate both early antibody-forming cells and germinal centers with hypermutation and memory in response to a T-dependent antigen. J Exp Med 198, 1923-1935.
Stein, J. V., and Nombela-Arrieta, C. (2005). Chemokine control of lymphocyte trafficking: a general overview. Immunology 116, 1-12.
Su, S. D., Ward, M. M., Apicella, M. A., and Ward, R. E. (1991). The primary B cell response to the O/core region of bacterial lipopolysaccharide is restricted to the Ly-1 lineage. J Immunol 146, 327-331.
Su, Y. W., Flemming, A., Wossning, T., Hobeika, E., Reth, M., and Jumaa, H. (2003). Identification of a pre-BCR lacking surrogate light chain. J Exp Med 198, 1699-1706.
Suzuki, H., Terauchi, Y., Fujiwara, M., Aizawa, S., Yazaki, Y., Kadowaki, T., and Koyasu, S. (1999). Xid-like immunodeficiency in mice with disruption of the p85alpha subunit of phosphoinositide 3-kinase. Science 283, 390-392.
Tarlinton, D., Radbruch, A., Hiepe, F., and Dorner, T. (2008). Plasma cell differentiation and survival. Curr Opin Immunol 20, 162-169.
Thompson, E. C., Cobb, B. S., Sabbattini, P., Meixlsperger, S., Parelho, V., Liberg, D., Taylor, B., Dillon, N., Georgopoulos, K., Jumaa, H., et al. (2007). Ikaros DNA-binding proteins as integral components of B cell developmental-stage-specific regulatory circuits. Immunity 26, 335-344.
Tiegs, S. L., Russell, D. M., and Nemazee, D. (1993). Receptor editing in self-reactive bone marrow B cells. J Exp Med 177, 1009-1020.
Tokoyoda, K., Hauser, A. E., Nakayama, T., and Radbruch, A. (2010). Organization of immunological memory by bone marrow stroma. Nat Rev Immunol 10, 193-200.
Tonegawa, S. (1983). Somatic generation of antibody diversity. Nature 302, 575-581.
Trageser, D., Iacobucci, I., Nahar, R., Duy, C., von Levetzow, G., Klemm, L., Park, E., Schuh, W., Gruber, T., Herzog, S., et al. (2009). Pre-B cell receptor-mediated cell cycle arrest in Philadelphia chromosome-positive acute lymphoblastic leukemia requires IKAROS function. J Exp Med 206, 1739-1753.

Tsubata, T., and Reth, M. (1990). The products of pre-B cell-specific genes (lambda 5 and VpreB) and the immunoglobulin mu chain form a complex that is transported onto the cell surface. J Exp Med 172, 973-976.
Tunyaplin, C., Shaffer, A. L., Angelin-Duclos, C. D., Yu, X., Staudt, L. M., and Calame, K. L. (2004). Direct repression of prdm1 by Bcl-6 inhibits plasmacytic differentiation. J Immunol 173, 1158-1165.
Tussiwand, R., Bosco, N., Ceredig, R., and Rolink, A. G. (2009). Tolerance checkpoints in B-cell development: Johnny B good. Eur J Immunol 39, 2317-2324.
Vale, C., and Botana, L. M. (2008). Marine toxins and the cytoskeleton: okadaic acid and dinophysistoxins. Febs J 275, 6060-6066.
Vecchi, M., Rudolph-Owen, L. A., Brown, C. L., Dempsey, P. J., and Carpenter, G. (1998). Tyrosine phosphorylation and proteolysis. Pervanadate-induced, metalloprotease-dependent cleavage of the ErbB-4 receptor and amphiregulin. J Biol Chem 273, 20589-20595.
Vettermann, C., and Jäck, H. M. (2010). The pre-B cell receptor: turning autoreactivity into self-defense. Trends Immunol 31, 176-183.
Vivanco, I., and Sawyers, C. L. (2002). The phosphatidylinositol 3-Kinase AKT pathway in human cancer. Nat Rev Cancer 2, 489-501.
von Boehmer, H., and Melchers, F. (2010). Checkpoints in lymphocyte development and autoimmune disease. Nat Immunol 11, 14-20.
Vora, K. A., Nichols, E., Porter, G., Cui, Y., Keohane, C. A., Hajdu, R., Hale, J., Neway, W., Zaller, D., and Mandala, S. (2005). Sphingosine 1-phosphate receptor agonist FTY720-phosphate causes marginal zone B cell displacement. J Leukoc Biol 78, 471-480.
Wagner, N., Lohler, J., Tedder, T. F., Rajewsky, K., Muller, W., and Steeber, D. A. (1998). L-selectin and beta7 integrin synergistically mediate lymphocyte migration to mesenteric lymph nodes. Eur J Immunol 28, 3832-3839.
Warner, N. L. (1974). Membrane immunoglobulins and antigen receptors on B and T lymphocytes. Adv Immunol 19, 67-216.
Weinreich, M. A., Takada, K., Skon, C., Reiner, S. L., Jameson, S. C., and Hogquist, K. A. (2009). KLF2 transcription-factor deficiency in T cells results in unrestrained cytokine production and upregulation of bystander chemokine receptors. Immunity 31, 122-130.
Werner, M., Hobeika, E., and Jumaa, H. (2010). Role of PI3K in the generation and survival of B cells. Immunol Rev 237, 55-71.
Willerford, D. M., Swat, W., and Alt, F. W. (1996). Developmental regulation of V(D)J recombination and lymphocyte differentiation. Curr Opin Genet Dev 6, 603-609.
Winkelmann, R., Sandrock, L., Porstner, M., Roth, E., Mathews, M., Hobeika, E., Reth, M., Kahn, M. L., Schuh, W., and Jäck, H. M. (2011). B cell homeostasis and plasma cell homing controlled by Kruppel-like factor 2. Proc Natl Acad Sci U S A 108, 710-715; Epub 2010 Dec 27.
Wittmann, J., Hol, E. M., and Jäck, H. M. (2006). hUPF2 silencing identifies physiologic substrates of mammalian nonsense-mediated mRNA decay. Mol Cell Biol 26, 1272-1287.
Wu, J., and Lingrel, J. B. (2004). KLF2 inhibits Jurkat T leukemia cell growth via upregulation of cyclin-dependent kinase inhibitor p21WAF1/CIP1. Oncogene 23, 8088-8096.
Wu, J., and Lingrel, J. B. (2005). Kruppel-like factor 2, a novel immediate-early transcriptional factor, regulates IL-2 expression in T lymphocyte activation. J Immunol 175, 3060-3066.
Yusuf, I., and Fruman, D. A. (2003). Regulation of quiescence in lymphocytes. Trends Immunol 24, 380-386.
Zhang, X., Srinivasan, S. V., and Lingrel, J. B. (2004). WWP1-dependent ubiquitination and degradation of the lung Kruppel-like factor, KLF2. Biochem Biophys Res Commun 316, 139-148.

I want morebooks!

Buy your books fast and straightforward online - at one of world's fastest growing online book stores! Environmentally sound due to Print-on-Demand technologies.

Buy your books online at
www.morebooks.shop

Kaufen Sie Ihre Bücher schnell und unkompliziert online – auf einer der am schnellsten wachsenden Buchhandelsplattformen weltweit! Dank Print-On-Demand umwelt- und ressourcenschonend produziert.

Bücher schneller online kaufen
www.morebooks.shop

KS OmniScriptum Publishing
Brivibas gatve 197
LV-1039 Riga, Latvia
Telefax +371 686 204 55

info@omniscriptum.com
www.omniscriptum.com

Printed by Books on Demand GmbH, Norderstedt / Germany